WINDOWS
ON
CREATIVITY
AND
INVENTION

BF
408

Jacques G. Richardson

Editor

. W56

Published with
the cooperation
of UNESCO.

Lomond
1988

Library of Congress Catalog Number: 86-083428

ISBN: 0-912338-57-1
ISBN: 0-912338-58-X (microfiche)

Several of the chapters comprising the present volume appeared originally in *Impact of Science on Society* 1984, No. 134/135 © by the United Nations Educational, Scientific and Cultural Organization, 7 place de Fontenoy, 75700 Paris, France. This material is reproduced with the authorization of Unesco. For other chapters, copyright as indicated.

Published by
Lomond Publications, Inc.
P.O. Box 88
Mt. Airy, Maryland 21771

Publisher's Preface

In one sense *Windows on Creativity and Invention* is an experiment. In another, it simply builds on the general theme that the intellectually and artistically creative process is complex, manifold and arguable.

The editor has assembled in 25 chapters the ideas of diverse, thoughtful people which are edifying in themselves. Together they contribute to better generic understanding which has value to all who are concerned with learning, doing, planning, inspiring, leading, organising, and refining the creative process and using the output.

The outcome from reading—and rereading—this book is likely to be quite different for each reader, a function of the background and readiness and openness which is brought to its study.

On the occasion of the publication of this, the fifth in this series of Unesco-generated books*, I express our appreciation for the contribution of Jacques G. Richardson, who in 1985 left the editorship of Unesco's outstanding journal, *Impact of Science on Society*. The breadth and depth of his understanding of science, his knowledge of the history of systems, and his insights into the effects of science and invention have been the key to his success in bringing the messages of outstanding scientists and interpreters of knowledge and its impact throughout the world into books related to salient themes—a journey most pleasant.

Lowell H. Hattery, Ph.D.
Publisher

*The five books are:

Integrated Technology Transfer
Models of Reality: Shaping Thought and Action
Violent Forces of Nature (edited by Robert H. Maybury)
Managing the Ocean: Resources, Research, Law
Windows on Creativity and Invention

Homo habilis

. . . Not only such arts as sea-faring and agriculture, city walls and laws, weapons, roads and clothing, but also . . . the amenities and refinements of life, songs, pictures and statues, artfully carved and polished, *all were taught gradually by usage* and the active mind's experience as men groped their way forward step by step.

Lucretius, Book V
(Cosmology and Sociology),
On the Nature of the Universe

I invent nothing. I rediscover.

Auguste Rodin

Creative minds always have been known to survive any kind of bad training.

Anna Freud

There are those who say we should not open our windows because open windows let in flies and other insects. They want the windows to stay closed so [that] we all expire from lack of air. But we say open the windows, breathe the fresh air, and at the same time fight the flies and insects.

Deng Xiaoping

TABLE OF CONTENTS

A few prefatory remarks serve to present the concepts of creativity and invention as human traits correlative to intelligence and initiative, their roles in the progress of culture and civilisation, and how all of these undergo ceaseless change and (for the most part) improvement in the welfare of mankind. Whereas innovation was once considered disruptive of the social order, today it is inevitable and moving along at a virtually exponential rate of speed.

Chapter 1

Introduction

Jacques G. Richardson

A science correspondent and former associate publisher of the French monthly journal La Recherche *('Research'), the author managed from 1973 until 1985 the Science and Society Section at Unesco, where he also edited the seven-language quarterly,* Impact of Science on Society. *Now an independent author and specialist in problems of technical communication, he can be reached at Cidex 400, Authon la Plaine, 91410 Dourdan (France).*

Imagination is nature's equal . . .

—Goethe

Intelligence, culture and civilisation

One of the reasons for throwing open some windows on creativity and invention is to honour the relationship inevitably to be found between the creative spirit and intelligence. There are other reasons too for opening these 'windows,' specified further along in this brief introductory chapter, throughout most of the sections that follow (especially in Chapter 24), and yet it is imperative that we deal first—if somewhat summarily—with the human quality that we call intelligence.

A suitable definition of intelligence for the purposes of this book is the one favoured by Robert Sternberg of Yale University, namely that 'intelligence consists of those mental functions purposively employed for . . . adaptation to and shaping and selection of real-world environments.'[1] By the time the reader will have finished with this volume, I hope that he or she will agree that our authors have indeed concentrated and elaborated upon such mental functions in describing how innovation occurs, where and when, and insofar as it is possible to speculate—why.

There may be a case, too, for drawing attention to creativity that is not primarily related to intelligence or education; this might be the situation, for example, in some of the oldest-tradition handicrafts. This dimension of the creative mode does not fit, however, the present collection of essays. Finally, in the book's penultimate chapter, we shall pay brief attention to some failed innovation because, mindful of C.G. Jung's admonition that 'knowledge rests not upon truth but upon error also,' I shall illustrate how thwarted success is part of the cumulative process of discovery and progress.

Charles Galton Darwin identified, in *The Next Million Years*,[2] the four major stages identifiable with the development to date of the human race:

- the discovery of fire
- the invention of agriculture
- the idea of living in cities (urbanization began some 6,000 years ago in Egypt, Iraq, China and Mexico)
- the scientific revolution.

No one can guess when *homo sapiens* mastered fire, but carbon dating and other modern methods reveal that the sedentary cultivation of edible plants originated about ten millennia ago. The so-called scientific revolution, much nearer our time, dates roughly from the works of Copernicus, Galileo, Descartes, Newton and Leibniz—or the first cultural evolutionary stage after the flowerings of the Chinese, Indian, Greek and Arab civilisations and the Italian Renaissance. This revolution now speeds ahead exponentially.

It is thus the progress made from the late 15th to late 20th centuries and onwards that concerns the authors of this book, throughout the social, natural and engineering sciences and in the humanities. Expressed in terms of geographic exploration, we are concerned with what man has been able to do constructively during the half-millennium since the four voyages of Christopher Columbus to the western hemisphere.

Despite the Polish, Italian, French, British and German names just cited, let the reader be assured from the outset that the authors writing in the pages which follow do not reflect only narrow Eurocentrism. It is the contrary that applies, and the reader will note that the very last contribution (the Postface) has been appropriately prepared by a citizen of a major country of the Pacific rim—the geographic region that is expected to become a leading global zone during the 21st century. Other authors from Asia and Latin America join forces with the European and North American contributors to round out the picture.

A silhouette of the chapters that follow

The volume has been divided, after this short overture, into four major 'movements.' In the first of these, four authors trace various innovative patterns related directly to the natural and exact sciences, what is commonly called *science* in English. Evgeny L. Feinberg establishes a relationship between innovative thinking processes and the challenge of habituation by mankind to novel theories, while Jean-François Doucet relates the imaginary, the symbolic and reality to the function of subjectivity in innovation. Robert Weisberg contests the conventional portrait of the stroke-of-genius operational method of the discoverer/inventor. Cesare Marchetti demonstrates an innovative method of analysing the total production of various artists and men of letters. These chapters comprise the rubric called The Natural Sciences.

There follows a division of the book designated Art, Music and Values, in which seven more authors take up the cues suggested by the section on science. Two artists trained in science explain their highly individualistic approaches to art forms that they deal with daily; the artists are Vladimir Bonačić and Alfred Kessler. Alfredo Planchart reviews cursorily the growth of knowledge and the rise of art—mainly in the

Occidental tradition; S.Y. Edgerton stresses the importance of the introduction of linear perspective in the art forms of the last five centuries; and Nils Wallin reminds of the pervasive companionship over the millennia between musical creation and mathematics. A young engineer, Gerard Blanc, recalls the vitality of introspection, vision and dreams that combine with awareness, memory, cumulative knowledge, self-organization and intensive work in the process of creation. This approach is complemented by that of Andrzej Wierzbicki, who examines

---

Creativity: causes and communication

Creativity . . . covers a constellation of different theoretical and practical, psychological and social [aspects]; it concerns economic, intellectual and artistic life, school and state . . . [Many] diverse causes are involved; conjunctural causes, of which a notable example is the 'Sputnik effect' in the United States, the birthplace of research into creativity, but also structural ones, in the sense that a *constant production of things novel* . . . seems to be a law of existence in modern societies—contrary to antiquity [when] time was a factor of disorder and novelty one of degeneration . . . [Such] different things as the enlarged reproduction of capital . . . or the avant-garde in art would appear to go hand in hand . . .

The existence of a variety of hypotheses and already solved problems, the possibility of effectuating recombinations constitute the pre-condition of invention. Their first institutional expression is to be found in an environment rich in the production and exchange of *information.*

        — Fernando Gil, Portuguese economist,
            in a paper presented at a Unesco seminar
            on creativity in today's society, held
            at Alcalá de Henares (Spain) in
            November 1982

---

the earthly behaviour of innovators as seen through the eyes of visitors from elsewhere in the galaxy.

The next major section is unusual because its authors concentrate on a specific cultural setting for novelty of many kinds: the Austro-Hungarian Empire during the final 150 years of its existence, and innovators born during this time (some of whom are still living). The editor underlines the transdisciplinary nature of this giant innovative laboratory and tries to explain some of the origins of the creative spirit in

Vienna at the end of a long era of political power, whereas Victor Weisskopf presents a charming personal reminiscence of the period. The major part of the book is more readily identifiable with the innovative processes with which the reader is familiar in contemporary life. In this sequence of chapters, Guy LeParquier, John Kao, Jean-Baptiste Donnet and Sir Bruce Williams examine in their separate fashions the challenges and responses inherent to the inventive spirit. Ouyang Wendao and Roy Uenishi weigh the significance of language and form recognition in the creative processes, while Djordjija Petkovski and the editor look at the significance of data processing (as well as the promise of artificial intelligence) and large-scale system problems. Wu Jisong compares and contrasts decision-making and management related to research in China and elsewhere.

Then, in the last section, the editor returns to review some changing philosophic patterns and loci of innovation, the influence of the sciences on evolving technology, and women as innovators. The final word is reserved for Junnosuke Kishida, whose concise essay explains how the processes of innovation themselves are changing—everywhere—because of the ambient characteristics of the research required, the technologies used in innovation, and how both of these affect society and culture.

Throughout the book, specific reference is made to examples of discovery, invention and innovative social processes, and the impulsions to be novel and to change (or else to prefer stasis) are treated. The factors of opportunity and chance are not overlooked, since order and randomness both have fundamental parts to play in evolutionary novation.

Linguistics inputs

For readers interested in the linguistic aspects of the compilation of a book of this kind, the following chapters are based on original texts in English: 1, 4, 5, 6, 7, 9, 10, 12, 13, 14, 16, 19, 20, 21, 23 and 24. Chapter 2 was written in Russian, Chapters 3 and 11 in French, Chapter 8 in Spanish, Chapter 18 in Chinese and the Postface in Japanese. Chapter 17 is based on an interview in English, and Chapters 15 and 22 are founded on interviews conducted in French.

Most of the translations were made by Unesco's centralized linguistic services—who, much too modestly, prefer to remain in anonymity because of the team nature of their effort. The editor coped as best he could with the rest. ■

Notes

1. R. Sternberg, Human Intelligence: The Model is the Message, *Science*, Vol. 230, 6 December 1985; Inside Intelligence, *American Scientist*, Vol. 74, No. 2, 1986.
2. C.G. Darwin, *The Next Million Years*, London, Rupert-Hart Davis, 1953, p. 91.

PART I

The Natural Sciences

*Creativity in art and science has tended to be accepted, throughout
historical evolution, as incomprehensible if not inane. Innovation
demands, however, habituation on the part of mankind before novel
theories and art forms can be accepted as part of the received culture
conforming with the popular ethic known as common sense.*

Chapter 2

Innovation's
own time machine

Evgeny L. Feinberg

*Evgeny Lvovich Feinberg is a physicist and corresponding member of the
Academy of Sciences of the USSR. As a departmental head at the
Academy's Institute of Physics, the author is concerned chiefly with
nuclear physics, radiophysics, acoustics, high-energy physics and cosmic
radiation. Among his more than 150 publications, his book* Kibernetika,
logika, iskusstvo (*Cybernetics, Logic, Art*) *was published in 1981 by*
Radio i Svyaz'. *Dr. Feinberg received a State Prize of the USSR, in 1983,
for his services to science.*

Is everything clear?

Before discussing innovation in science as a whole, it would be as well first to consider whether at the present time, in the twentieth century, science is not something special such that the problem of innovation should also be understood rather differently than in the past.

Ours is the age of atomic energy, space flights and computerization; the age of the theory of relativity and quantum mechanics; the age of genetic engineering and the robotization of industry; the age of new synthetic materials and transistors, of the green revolution in agriculture, and of antibiotics. It is an age of staggering scientific and technological achievements, unprecedented, unheard of and mostly even unforeseen. Is this not a miracle in the long history of mankind? It has already been dubbed the age of the scientific and technological revolution, the start of which has even been precisely indicated as the late 1940s. Everything is clear. Everything?

First, what sort of a revolution is this that has been going on for half a century already and whose end is not in sight? By revolution, as distinct from evolution, one usually in fact understands something fast-moving and explosive, overturning what was previously customary and stable, something like a new frontier.

Second, awed admiration of the science and technology of one's time is by no means peculiar to our age. This is the way it has always been, or at least for the last four centuries. If proof of this is still needed, we shall produce it now.

We must, therefore, give careful thought to what the specific feature of contemporary science is and to what the scientific and technological revolution really comprises.

A word first about its place in history. Was the development of science in itself in the nineteenth century, for instance, any less revolutionary than in the twentieth century? Clearly not. Any other conclusion disregards the historical perspective. Let us take as an actual example physics, with which I am familiar.

The emergence of modern physics

In the twentieth century, physics has penetrated into the microcosm, the atom and even into its nucleus. This, however, was only a subsequent stage in a single, protracted process. In fact in the nineteenth century, from Dalton in the early years to Maxwell and Boltzmann in the latter part, the process of establishing the atomistic conception of the structure of matter and the closely associated kinetic theory of heat was going on all the time. At the beginning of the century, it was believed that the thermogen was a fluid which is also heat, but the century ended with the consummation of the supreme, all-embracing science of thermodynamics and its statistical interpretation. The mid-nineteenth century saw confirmation of the law of conservation and transformation of energy, which together with the second principle of thermodynamics became a powerful curb on any fantastic and innovatory imaginings. Finally, electricity and magnetism developed in an unprecedented way and were combined by Maxwell into a single orderly theory that laid the basis for the entire field of electrical engineering, with electric motors, electric lighting and heating, the telephone, telegraph and the boundless realm of radio engineering.

Meanwhile, in the early nineteenth century, the French scientist and author of a physics course, Abbé Hauy,* in his admiration for the achievements of science in the eighteenth century, wrote: 'The study of electricity, enhanced by the work of so many renowned physicists, seems to have reached a limit beyond which there can be no marked progress, so that the only hope left to workers in this area will be to corroborate the discoveries of their predecessors . . . It is as if science is reaching a static state.' This, we should remember, was written when the entire world used candles, lamps or chippings for lighting and when the magnetic effect of current was not known—meaning that there was no electrical engineering or radio technology, no electric telegraph, no dynamo and not even electric toasters. 'At that time, the convulsive movements observed by Galvani in the muscles of a frog attracted the astounded attention of physicists.' Volta discovered the occurrence of differences in potential when different metals are brought into mechanical contact, 'and this fact was transformed in his hands into the embryo as it were of an amazing device that, through its fruitful results, has come to occupy a foremost place among all the apparatus that human genius has produced for physics' (although in the modern study of electricity this device occupies one of the lowliest places).

The awesome and the awestruck

It was the same in earlier years as well. There was nothing to compare, for example, with the admiration shown for the achievement of science a hundred years previously, in the early eighteenth century, when Newton first provided a mathematical theory for physical phenomena and when Leibniz, Descartes and Huygens were completing their work, when

---

* René Just (1743-1822), founder of crystallography.

differential calculus was devised, the major laws of optics worked out, and so forth. The awe felt by the educated public at the achievements of science was universal.

The same was true a hundred years earlier still, however: in 1611, the prominent English poet and churchman, John Donne, wrote the poem 'An Anatomie of the World' which includes these lines:

'For of Meridians, and Parallels,
Man hath weav'd out a net, and this net throwne
Upon the Heavens, and now they are his owne'.

This was a tribute to the resounding successes in astronomy of Copernicus, Kepler and Galileo. The admiration expressed in those lines written nearly four centuries ago is comparable only to our wonderment today at space flights.

Thus the successes of science in the twentieth century are in themselves a natural extension of its advances in previous centuries, while the striking impression they make on our contemporaries is a natural reaction of four centuries' standing.

Our modern scientific and technological revolution cannot therefore be seen merely in terms of new scientific achievements since here there is nothing fundamentally unusual.

Perhaps the significance is in the practical, technical applications of science. A historical digression will readily convince us that here, too, there is nothing basically new. The technical applications of science in the seventeenth and eighteenth centuries—ocean-going navigation, medicine (with drugs and the thermometer), dyestuffs in the textile industry, the steam engine, explosives, metallurgy, hydraulic engineering of every kind and nascent electrical technology—in short, the whole range of technology that distinguishes man's living conditions and activity in the late nineteenth century from the situation in Shakespeare's day—stems from the use of scientific achievements. In this respect, nothing striking or unusual has affected the historical process in our day.

If it is decided to call what has happened in the twentieth century in science and technology a revolution, then this is a stage of 'permanent revolution' lasting many hundreds of years and innovation here is traditional.

What there is really new

All the same, there are new things, but these are to be found in the mutual relationship of science and human society.

The material indicators of scientific activity—expenditure on science, the number of people engaged in 'producing science' and the number of scientific books and papers—have been extensively studied in quantitative terms, using the material of the last few centuries. We know that all these indicators have in general grown exponentially by a number of percentage points a year. It can be assumed, for instance, that these indicators double every fifteen to thirty years, which means an approximate annual growth rate of 2.5 to 5 percent. This law of growth helps us to understand at the quantitative level why, as has been shown above, at all times and at all stages of the development of science, people have reckoned the scientific successes of their own day to be unprecedented and particularly far-reaching, and why they have been so proud and admiring of them.

The point is that they were indeed right on each occasion. The exponential law of growth possesses one purely mathematical feature: Whatever the period taken as coeval with doubled growth (fifteen to thirty years, as we have approximately assumed), it will turn out that during the period in question as many books are published, as many scientific papers are written and as many scientific workers are trained as in the entire previous history of mankind. It is no surprise then that each person is struck by the fact that during his lifetime the scientific and technological progress made surpasses man's entire previous achievements. It was always so, however; and the same could be said by the contemporaries of Pushkin, of Swift and of Molière. And by you and me.

Nowadays up to 4 percent of national income is spent on science in the industrialized countries. Meanwhile, annual population growth in the developed countries stands at a mere 0.2 to 1 percent. Only in the developing countries, on account of increased life expectancy after the lean colonial times, does the birth rate reach 2 to 3 percent. In comparing these figures, we arrive at the answer to what the specific feature of present-day science is. While in earlier times science occupied a negligible proportion of the population and expenditure on it represented a minute slice of the national income, the situation has radically changed. The previously insignificant proportion of the population engaged in science is now by no means so small that it can be disregarded and science has become one of the broadly practiced occupations. There can hardly be any further increase in that proportion, and the number of people engaged in science cannot grow at the previous rate. In the industrialized countries, this growth rate can no longer exceed the population growth rate. The proportion of national income spent on science cannot increase at the previous rate either, since it, too, is already very substantial. Both these indicators have reached a certain saturation point. This is mainly what the scientific and technological revolution has been about, and nowadays innovative science and technology have entered every home.

The new meanings of science

Science has become a direct productive force of society, and in the capitalist countries it has become just about the most profitable business. We now commonly mingle with scientists, with the creators of new technology, with their assistants and with those who back up their activity (publishing books and journals, making laboratory equipment, and so on). Science is made not by little-understood 'recluses' about whom nothing may be known for a lifetime; scientific and technological creation are fully visible to all. When I was preparing to enter the physics faculty more than half a century ago, I was sometimes asked: 'Does this mean you're going to do physical culture and sport?' Nowadays such a reaction would be impossible, even in the remotest backwater.

The notion of scientific and technological revolution is to be seen in this change in the position within society of scientific and technological activity. In science itself, taken on its own, it is hard to discern any particular revolution.

If we concur in this, we must conclude that the problem of innovation in science should not be regarded as something really peculiar to our times. Innovation is an essential component of science, and has occupied an identical position throughout at least the last four or five centuries. Science is advanced by previously unknown and uncomprehended breakthroughs, and I insist that this has always been so.

The only specific feature about which one can still speak is that innovation is being freed from the fear of the 'ideological interdictions' which in the form of various dogmas were previously so strong and weighed so tragically on Copernicus and Galileo, or which, in the form of scholastic Aristotelian norms, hindered the innovation of many. A case in point is Kepler. As we know, Kepler reckoned that the orbits of heavenly bodies must be ideal, meaning circular. He discovered the elliptical trajectories of planets quite against his convictions.

Nowadays such norms are commonly known as governing paradigms. Even now, such paradigms certainly limit innovation in cases where in order to move forward one has to break away from them and switch to a new paradigm. But there, pressure is now not so great.

The emergence of 'ideological interdictions' in particular dramatic situations is nowadays always of a temporary character. After curbing the development of a specific branch of science for a time, the interdictions are overcome because the successful development of science is a vital necessity to society as a whole and to the progressive development of mankind. This in turn has been determined by precisely that (a) change in the relationships between science and society and (b) penetration of science into the entire life and activity of society, which we have outlined above as the main ingredient of the notion of scientific and technological revolution.

Where innovation in physics is concerned, we can mention and discuss yet another widely held opinion which is evidently fallacious when viewed in an historical light. This is the frequent claim that, as we penetrate into ever more complex issues relating to the structure of matter and to the universe and space, new ideas become increasingly abstract and unclear. The usual examples advanced are the corpuscular-wave dualism of particles in the microcosm and the expanding and nevertheless boundless universe of the general theory of relativity.

The absurdity factor in scientific evolution

When innovatory notions of this kind emerge in science, they are indeed very often not immediately clear, go against 'common sense' and are perceived as a dubious mathematical abstraction. It is easy to see, however, that this too is by no means a new situation. Just about every fundamentally important advance in physics and astronomy *that was a genuine innovation* was initially perceived as contradicting common sense, incomprehensible and not at all evident. This is simply because any new, fundamental discovery always represents a departure from customary experience assimilated by the consciousness in the form of either everyday, routine common sense or a settled scientific system, or both.

The switch from a geocentric system of the world to a heliocentric system thus seemed, at the time, absurd because everyone could clearly see that the sun rose and set. Even the roundness of the earth was dismissed out of hand simply by drawing a globe and two human figures at the antipodes, standing feet to feet at opposite ends of the earth's diameter. The 'absurdity' of the situation was obvious. The later affirmation that the atmosphere presses upon man with a huge force—one kilogram per square centimetre of the body's surface—and yet does not crush him was equally incomprehensible. And even Otto von Guericke's experiment with the Magdeburg hemispheres, though constituting scientific proof, was hardly able to dispel all the doubts prompted by 'common sense'.

Such examples can be enumerated *ad infinitum* right up to our own day. What is important is that as a rule, such discoveries initially seem 'mathematically true' but quite incomprehensible. But as new experience is assimilated in science and in everyday life, the fog lifts and a new common sense always emerges. Clarity is restored. In short, the loss of clarity with each innovative breakthrough in science is made good after the assimilation of a new realm of facts and mastery of the theoretical and the empirical. Such loss is always temporary. Let us remember what we ourselves have witnessed with the 'incomprehensible', 'formal mathematical' conclusion of the theory of relativity regarding the slowing down of all processes, including biological ones, which takes place when a body moves at very high speed.

Time compression, à la Einstein

Nowadays a favourite theme of science-fiction writers is that of the person who retains his youth after setting off on a long space journey and returning to earth to meet his distant descendants who in the interval have reached advanced old age. This subject is based on an undoubtedly true scientific conclusion from the theory of relativity. It is reliably confirmed by experiments with inanimate objects. For example, when an extremely high-energy particle collides under cosmic radiation with the nucleus of an atom of air, this produces a pair of unstable mesons, one of which may be high-velocity and other low-velocity. The former moves further without 'dying', taking a very long time to disintegrate, while its slower 'twin' very soon breaks up into new particles which in turn are transformed into other particles, and these distant descendants of one of the 'twins' will coexist with the faster-moving 'young' brother of their ancestor.

Although the subject has a firm scientific foundation, it is nonetheless astonishing. All the same, however much the reader of such a novel may smile and shake his head in doubt, he already knows that this really is a possible story in terms of the new common sense. Each similar situation with innovation contradicting common sense, when clarity is temporarily lost, is a natural and inevitable stage in the development of science.

A reservation is needed here. Clarity used to be understood as the reduction of a phenomenon to a mechanical model with the displacement of particular bodies, their elastic variations, and so forth. But this narrow point of view must be discarded since we now know, for instance, that it is not elastic phenomena in a hypothetical ether that underlie electromagnetic processes. On the contrary, the elastic properties of bodies are to be explained by the electromagnetic interactions of their component electrons and atomic nuclei.

A few concluding observations

Let us try to sum up. Innovation in science has at all times meant the mind breaking free from fixed ideas and studied facts and phenomena. At all times, innovation—where fundamental notions were concerned— was liable to be accompanied (and usually was accompanied) by a conflict with common sense, which is always a consolidation of facts that have been previously well studied empirically and theoretically. And all this is equally applicable to innovation in art.

Of the most audacious innovations in science, not one is arbitrary. Innovations 'sense' and take account of the restrictive framework of those general principles—previously established and that must remain correct (and for the interpretation of which settled theories were established in the first place).

Such innovation is always effected as an intuitive act going beyond the framework of the previous system of logic and asserting new axiomatic foundations of theory. They are verified by the fact that their conclusions coincide with experience and practice, and if no contradiction is found, these new axiomatic foundations are recognized as true. As we know, however, the criterion of practice is limited and incomplete. Therefore the very statement that something is 'in keeping with experience'—which is always finite and limited—is an extra-logical, intuitive judgement that is not absolutely true. Only in this way is the development of science possible. After an extension of the area of activity under investigation, the limited truthfulness of knowledge may appear and a new change of view and new extension of the axiomatic base will occur.

Innovative art is also never entirely arbitrary. It is always to some extent restricted by the previously 'settled' art, which it transforms. One of the bold innovators in music, Igor Stravinsky, wrote: 'The necessity of limitation, of voluntarily accepted restraint, originates in the depths of our very nature and touches not only the realm of art but all conscious manifestations of human activity. This is the requirement of order, without which nothing can be created. Yet any such order involves constraint. It would be but vain to see in this an obstacle to freedom. On the contrary, restraint and limitation assist the burgeoning of that freedom and simply prevent it from degenerating into total license. It is precisely in this way that, in adopting a ready-made, consecrated form, the creative artist is in no way constricted by this when it comes to displaying his own individuality. I would even say that individuality stands out more clearly and in greater relief when it can be exercised within conventional and clearly demarcated bounds'.

In art as in science

Thus the innovator in art as well, in setting aside old-established principles, standards and rules, does not reject them totally. He may create a new 'system of logic' that may initially (though this certainly is by no means inevitable as in science) seem absurd and at variance with established common sense. 'Genius is talent which lays down its own rules', said Kant. But if the innovative artist has correctly felt the incipient spiritual evolution of his public and has properly understood what, in the existing and settled, must be retained as something of undying value, then the subsequent test of experience will confirm the authenticity and fruitfulness of his new principles and new axiomatic bases. This authenticity will show itself in the fact that his art will enter into people's extended common sense and will cease to be, as was said in the case of science, incomprehensible and unclear.

Art that is initially devoid of common sense, looks like nothing on earth and seems quite arbitrarily constructed, subsequently reveals clear and profound links with the preceding period in that art. It absorbs a great deal from the logic of the previous art, and the combination of what has been intuitively discovered with the logically regulated becomes evident.

Happy indeed is the artist if he follows in the footsteps of Shostakovich and Stravinsky, for instance, who succeeded in their lifetime in instilling in people a new common sense in music.

True, combining the logical and the intuitive in science is not the same thing as in art, and the basic purposes of science and art differ. When it comes to innovation and creativity, however, and the way in which they tie in with what is habitual, traditionally established and accepted (yet already inadequate in terms of new scientific, artistic or fundamental experience), we cannot fail to observe much that is strikingly similar in science and in art. ▄

Scientific objectivity is often depicted as being a guarantee of the truths it sets forth. Today, far-reaching epistemological changes, extending far beyond apparent certainties, have made it possible to take subjectivity into consideration. Located between the Imaginary of scientists and the Reality which is the focus of their observations, the Symbolic must now be used as a means through which this subjectivity may be expressed and shared.

Chapter 3

Scientific objectivity
is subjectivity shared

Jean-François Doucet

Jean-François Doucet is an engineer and senior research worker at the University of Oslo. After a basic science course he became interested in languages as used within their cultural context (he has studied German in Mainz, Swedish in Dalecarlia and Scania, Danish in Copenhagen, and Norwegian in Oslo). As head of the science and technology information services of ENSIA (a French school of engineering), he was required to create and organize data bases and data banks. His study of the use of scientific and technical information for creative purposes involved him in theoretical work on metaphoro-metonymic alternation, seen as one of the keys to human creativity. A few years ago, he founded the Creativity Science Institute, which is associated with the University of Oslo, with the help of interested organizations such as the Organisation for Economic Co-operation and Development (OECD) and other foreign institutes. Jean-François Doucet published an article focusing on words in the issue of Impact *devoted to models (Vol. 31, No. 4, 1981). He may be contacted at the following addresses: G. Vigelandsv. 46, Oslo 2 (Norway), or 26 rue A. Lemoine, 95300 Pontoise (France).*

Science has patiently worked towards replacing the tangible world with an intelligible sum of knowledge. [1] It develops its own methods, using thought to move forward from a subjective observation to a rational representation of observed phenomena. Beginning with scattered impressions, it builds up a front of truths whose validity is assessed through the effective correspondence of science to reality.

This once-flawless facade is now beginning to crack. Science is being required to furnish explanations and has become the target of sharp criticism which strikes at its very foundations. Gaps have been opened in modern knowledge, whose already explosive development has been the battleground of many a revolution. [2] In many respects, these transformations have been to science what abstract painting was to representational art: a bombshell with more than the obvious immediate effects.

Normative science, which presumed to dictate what was 'real', has clearly had its day. [3] Other forms of intelligibility are being seen as so many possibilities to be investigated. [4] In future, in order to preserve the objectivity of its observations, science will have to take account of the subjective factor attached to the observer. Science admits of more than one kind of relationship between the object and the observer, identifying three categories. The three terms: the Real, the Imaginary, and the Symbolic should be understood in the sense of Lacan, the contemporary French anti-psychiatrist, as I explain below. *

Three categories

The Real is the void; a space devoid of human factors, in which scientific phenomena have their being. A pictorial representation—the imagination—is stored in the Imaginary, which is unconscious in part, but whose conscious part perceives its manifestations. [5] These representations, taken as a whole, are contained in the Symbolic, [6] whose signs come into circulation and form what is commonly termed an understanding of facts, persons and things. This flow of information gives rise to a set of scientific symbols generally accepted by individuals in a position to know.

---

* Jacques Lacan (1901-1981) was the first in France to advocate anti- or counter-psychiatry. The terms Real, Imaginary and Symbolic are used here in the sense in which he intended.

Scientific creativity

Creativity, in this context, is precisely this capacity for symbolic re-arrangement. If compared to a piece of cloth, the Symbolic is stained with factors produced in the Imaginary, which penetrate both warp and woof. In this way, the cloth is gradually transformed. The Real thus appears along two axes:

The *syntagmatic axis*, like the warp of our piece of cloth, through which pass the threads of the woof. On the Symbolic plane, these threads represent all the relationships existing between elements arranged in a structure similar to that of a sentence. The syntagmatic axis provides explanations which link cause and effect. For example, there was a syntagmatic relationship between Koch's bacillus and tuberculosis, in the form of experimental descriptions. [7]

These relationships, like the woof of our fabric, become implicit along the *paradigmatic axis*. Using the preceding example, tuberculosis was implicitly linked to Koch's bacillus from the day the scientific community decided it was responsible for this disease.

In a more general sense, all scientific progress can be described as the shifting of knowledge [8] from the syntagmatic axis, around which theories are formed (sometimes unexpectedly) to the paradigmatic axis. Creativity may be distinguished from progress by a dissociation between the syntagmatic warp and the paradigmatic woof, which precludes any continuity or deduction from one state of knowledge to another. This dissociation is known as an epistemological gap. [9]

The cause of this gap is to be sought in the Imaginary (hence Einstein imagined an observer travelling on a light-ray), wherein, before appearing on the symbolic plane, the warp is formed by a paradigmatic dissociation, or leap. After images of the new reality to be explained come the words and the systems into which they are organized, thus relieving them of their emotional weight, whether of envy or anxiety, and making them capable of communication by the Symbolic.

Shared subjectivity

Scientific truths in this context have an extremely subjective origin: the Imaginary. Their objectivity is only real insofar as they have been critically tested against other subjective phenomena. But they are never pre-determined or completely unchanging. These truths merely form a reliable basis for further activity by various parties. They are never an absolute guarantee of an exact correspondence between the reality of a phenomenon, the explanation that science gives of it, and how one

---

KREATEK: an interdisciplinary data bank for creative science

A large number of works having focussed on creative science during the last thirty years, the need was felt for a computer-assembled and processed documentation system in this field. The University of Oslo, welcoming all the various manifestations of this relatively new branch of knowledge, made possible the establishment of KREATEK. The documents stored in KREATEK, which combine the history, philosophy and psychology of science (books, magazines, reports, theses, microfilms, etc.) are a practical demonstration of the interdisciplinary field of possibilities now open to creative science. Its internal coherence relies on four factors: its fields of application in the arts, technology and science are covered by its preoccupation with the phenomenon of creation, and languages covering the forms of creation characteristic of each discipline make it possible to clarify both the methods and the specific objects of creativity. (See Fig. 1.)

Through KREATEK, creative science can be of invaluable assistance to researchers, artists and engineers interested in methodical and generative innovation processes. On request, KREATEK's data bank will furnish photocopies of texts answering questions in specific fields.

In addition, given the profusion of scientific and technological information available today, KREATEK makes possible the creative use of a resource produced by the exponential growth in knowledge. The separate concepts of information and scientific and technological creativity are thus productively brought together by KREATEK.

---

imagines it to be. Knowledge, in this respect, functions much like a system of symbols which is held up against reality, without its being clear whether this system is the only suitable one, or whether the system corresponds exactly to the explanation of reality it is required to give. In any event, science must always base its assertions on some element of doubt; it is as if a tear in the fabric of knowledge could occur at any time, thus bringing us that much closer to reality when the fabric has been rewoven.

These truths are discovered like statues unveiled at an inaugural ceremony, only to be covered up once again with a fabric of new explanations. Science develops as if the Imagination were holding a screen between the reality which it contains and the reality which, on the other side, still remains somewhat obscure. As the beauty, cruelty and

INTERIORIZATION

Desire for an object or a person, here and now

Desires (psychoanalysis)
Dreams
Rites
Symbols (psychology)

Investment in spatial and temporal movement

Emotions
Imagination, ideas associated with images

EXTERIORIZATION

Representation for others, elsewhere, before or after

Signs (semiology, semiotics)
Written language (literature, linguistics)
Images (painting, sculpture, architecture)
Notes (music)

Interpretation of the object over time, extensively, everywhere

	Philosophy	Psychology	History
Arts			
Techniques, technology (innovation)			
Science (history, epistemology, etc.)			

Explanation of the total, universal object

Myths
Stories and legends
Religions

Fig. 1. Classification of KREATEK documents for creative science.

harshness of reality are unendurable, reality must be veiled by a theoretical approach. Truth, on the other hand, is blinding in its clarity. As a result no one can really grasp the truth and explain it. It is an ultimate condition like death. Without ever really attaining truth, science endeavours to explain it. Truth, like death, is thus a belief—incomplete and unknown—which continually needs to be demonstrated by constant testing. ■

	Mini-glossary
Metaphor (= symptom)	Figure of rhetoric using analytical substitution to denote an action or object uncommon to either: 'He used his heaviest artillery to win the debate.' In psychoanalytic theory, designates by analogy a subconscious mechanism of rejection.
Metonym(y) (= desire)	A figure of rhetoric using the name of an object closely associated with another as its substitute: 'I read Aeschylus [i.e. the works of Aeschylus] the entire day.' In psychoanalytic theory, designates by analogy a subconscious mechanism of shifting or displacement.
Paradigm	A model or simulation of a thought, an object, a pattern or a system.
Syntagma	Like the word 'syntax', from the Greek verb '*syntassein*', to arrange systematically; a term in linguistics, also meaning an orderly collection of thoughts or writings.

Notes

1. S. Weil, *Sur la science*, Paris, Gallimard, 1966.
2. T. Kuhn, *The Structure of Scientific Revolutions*, Chicago, University of Chicago Press, 1975.
3. P. Scheurer, *Révolutions de la science et permanence du réel*, Paris, Presses Universitaires de France, 1979, p. 338.
4. Ibid., pp. 249-69.
5. O. Mannoni, *Clefs pour l'Imaginaire*, Paris, Le Seuil, 1979, p. 319.
6. G. Rosalato, *Essais sur le symbolique*, Paris, Gallimard, 1969.
7. J.-C. Gardin, *Le Syntol* (*Syntagmatic Organization Language*), Brussels, Euratom, 1964, p. 10.
8. This paradigmatic/syntagmatic opposition is used by Saussure in his description of associative/syntagmatic relations, and by Jakobsen in his syntactic/semantic differentiation.
9. This oppostion is based on two figures of speech (metaphor and metonymy), and may be compared to the metaphoro-metonymic alternation described in Rosolato, op. cit.

Early analyses of the generation and spread of innovation led the author to examine their generator and scatterer, the human being. Applying the same analytical techniques as those used in the examination of social structures, he found that homo sapiens is highly ordered and regulated and that his output can be simply described and, to a certain extent, predicted. This becomes a generalization of the principle that the biological paradigm can be extended to social behaviour.

Chapter 4

Action curves
and clockwork geniuses

Cesare Marchetti

Dr. Marchetti studied physics at the University of Pisa and the Scuola Normale in the same city. He worked on problems of heavy-water technology in Milan and Buenos Aires and on applied surface physics at the Battelle Institute in Geneva. A former head of division at AGIP Nucleare and the European Community Research Centre, the author joined the International Institute of Applied Systems Analysis (IIASA) in 1974, where he is a research scholar. In 1979, he was awarded an honorary doctorate in science by the University of Strathclyde in Glasgow. Address: IIASA, Schlossplatz 1, 2361 Laxenburg (Austria).

Tu ne quaerisis, scire nefas,
Quem mihi quem tibi
Finem dii dederint, Leuconoe, nec Babylonios
Temptaris numeros.

Horace, *Odes*, I, xi, 1*

Fate and destiny are cross-cultural elements in philosophy, religion and
art, perhaps reaching the highest emotional and intellectual density in
Greek tragedy and their most subtle expression in Calvinist theology.

Man perceives external forces, endowed with an internal logic, that
frame his actions, feelings and thoughts in an unbreakable (if not visible)
cage. It would surely be strange if such a general perception did not
correspond to some objective fact.

That the society around us frames our actions is a trivial observation.
Yet I conclude, from my own socioeconomic analyses,[1-3] that this
framework is quantitative and that the underlying logic is hard.
Although these factors of framework and logic have been expressed
previously at the collective level, it may be possible to accommodate the
performance of the individual. Just as Greek tragedy and Calvinist
theology focus on the individual, I have also confined the present analysis
to the individual in order to determine if there is a strict logic operating
in this sphere.

Logic certainly abounds as long as we remain within the biological
realm. The more that we know about genetics, for instance, the more we
can observe its pervasive (if not steely) control. Indeed, genetics may even
clock the very day of our death, and evidence to this effect is mounting
up.[4] The effects of time and biological tempo on intellectual activity are
a good example of what I mean. Although patterns have been sought and
found,[5] basically these have not gone beyond the level of qualitative
description. Youngsters are 'naive' and oldsters are 'tired,' so that
activities such as sports and the writing of books are functions of one's
age.

The methodology is explained

I have sought to harden the analysis and, as usual, I began with hard
and proven facts. Also as usual, I searched extensively for patterns and I

---

* The first line translates: Pray, ask not—such knowledge is not for us.
I leave to the reader the challenge of working out the rest. (See p. 38.)

avoided as much as possible entanglement on the flypaper of explanations.

As an appropriate documentary base, my method was to concentrate on persons whose actions have been adequately studied, classified and appreciated. Artists and scientists provided a good starting point. I assumed that a piece of art or science is the final expression of a *pulse of action*, one that began in some obscure part of the brain and then worked its way through all the intermediate steps to a coherent action.

This led me, in turn, to study the time patterns of these pulses because, in my heuristic mental image, actions are transcodifications of information structures. The relationship between pulses and time patterns is true in the biological dimension, where information is processed in DNA, as well as in the personal or social dimension where the coding involved has a different substrate but where the same, basic processing rules apply.[6]

One of the consequences of this mental imaging is the assemblage of pulses of action into growth processes. Integrating these over time, I found that the growth function is in fact the temporal organizer of these processes. The curves appearing on the graphs that follow presuppose a final target or niche and a rate which, in biology, is directly proportional to the product of the level reached (i.e., the current size of a given population) and the part yet to be bridged (i.e., the unfilled niche) in order to meet the target. Because competition, for example, can keep one from reaching this final target, I chose to call this objective a virtual or *perceived target*. This element is a constant in the growth equation and remains so throughout the entire growth period.

I have applied these concepts to analyses of a disparate array of socioeconomic structures, classifying the growth of populations—the term is used here in the statistical sense—of railway networks (with the inauguration of the first rail line taken as the birth date),[1] automobiles,[2] and clusters of innovations.[3] Application to individuals then comes automatically, provided that one takes the backward step of assuming that information is processed essentially at the individual level, as is the case in society as a whole. The reasoning is, at any rate, heuristic, and the results will tell if we are on the right path.

The three-parameter logistic curve

The analytical results pertaining to a selection of outstanding personalities from the worlds of science, music and the visual arts are given in Figures 1 to 13. In each case I took as the population the 'brain children' of these specialists, plotting their growth and then fitting the data to a logistic curve.* The resulting chart is what I call the *action curve*, analogous to the growth curve in the case of an individual. I refer to the saturation level of an individual's output as the *perceived potential.*

It will be noted that these logistics have only three parameters. The time distribution of the work of a great creative spirit can thus be condensed into three numbers. The first defines the perceived potential; the second fixes a rate constant—or, more intuitively, a time constant— that measures the spread over time; and the third is a time cursor that locates the structure within the lifespan of the person in question. These parameters can be extracted from a segment of the data, making the rest of the data determinate.

It should be obvious that I did not take direct aim at the holy cow of creativity, only at its mooing, as it were. It is for this reason that I call the charts *action curves.* And because I believe that this relationship is of general character, I am continuing the search for appropriate indicators and statistics in order to prove that the curves apply as well to less noble deeds (e.g., criminal acts).

These three numbers (perceived potential, rate constant, time cursor) create possibilities for prying into the life of a genius, and categorizing and classifying it. To illustrate further what I mean by the perceived potential, consider the amount of beans one carries in his bag and the amount remaining when one dies. Looking at the cases used here (as well as others that have been excluded from my account so as to keep the paper brief), I find that the beans left over come to 5-10 percent of the total potential. In other words, when Mozart died at the age of 35, he had already said largely what he had to say (see Figure 12).

An admission of possible biases

My choice of cases may be biased, admittedly, since the choice was determined by the availability of serious 'data strings.' My personal interest in the visual arts, for example, and the abundance of critical information to be found in my personal library prejudiced me towards the painters.

As my investigations in this field are still at the beginning, many questions remain open, e.g., how large is the area of application? My guess is that the observed pattern is, indeed, a general one that can be

---

* A logistic curve represents a function related to such things as population growth, rise in the number of laws passed or inventions made. See Figures 1-14.

used to describe human action. As a counterpart to looking at the brain children already mentioned, I show the results in Figure 14 of an analysis of (non-symbolic) childbearing. The chart reveals some interesting peculiarities about the social and cultural modulations of human fertility.

Elsewhere, I have examined the generation and spread of innovation in Western society.[7] Mankind is, after all, the prime mover and final recipient of innovation.

Comments on the charts

Figures 1 to 14 correspond to the cumulative number of objects produced, independently of their size and of the importance attributed to them. The fitting equation, in all cases, is a three-parameter logistic of the type

$$N(t) = \frac{\overline{N}}{1 + \exp - (at + b)}$$

where N is the cumulative number of objects at time t and $\overline{N}$ is the saturation level or asymptote. $\overline{N}$, a and b have to be calculated by 'fitting' the data. If we define F as

$$\frac{N(t)}{\overline{N}} , \text{ then } \log(\frac{F}{1 - F}) = at + b,$$

a linear form.

$\overline{N}$ is given in parentheses; it is the perceived potential referred to earlier. ΔT gives the temporal spread of production, or the time to go from 10 percent to 90 percent of $\overline{N}$; it represents a more intuitive way of expressing a. The parameter b is a time cursor locating the pulse within the life of the person portrayed.

The year of maximal productivity is indicated in most cases, and it corresponds to the point in the curve where $0.5\overline{N}$ is reached. The first data points usually appear below the equation line, a fact that I interpret as catch-up. In other words, the creative impulses existed before they could be technically expressed—as is the case in Figure 14 with the fertility of girls aged 14 years. Yet as the dynamic of catching up shows, the impulses are delayed rather than suppressed. Girls have already met their prescribed 'quota' by the age of 18, as do artists very early during their creative period.

As death approaches (very often around 90 percent of $\overline{N}$), there is usually a slight increase in output with respect to the equation's prescription. Mozart's productivity is typical. In the case of Shakespeare, the 'extra' play was in fact written (as I discovered later) by someone else.

Postscript

Other cases than the ones shown here are currently under elaboration. These show that sometimes a double life of creativity, one with two pulses manifesting themselves, may be possible.   ■

Notes

1. C. Marchetti, Society as a Learning System: Discovery, Invention, and Innovation Cycles Revisited, *Technol. Forecasting and Social Change*, Vol. 18, 1980, p. 267.
2. ————, *On A Fifty Years' Pulsation in Human Affairs: Analysis of Some Physical Indicators* (publ. PP-83-5), Laxenburg, IIASA, 1983.
3. ————, The Automobile in a System Context: The Past 80 Years and the Next 20 Years, *Technol. Forecasting and Social Change*, Vol. 23, 1983, p. 3.
4. A.T. Winfree, *The Geometry of Biological Time*, New York-Berlin, Springer Verlag, 1980.
5. D.K. Simonton, *Genius, Creativity and Leadership*, Cambridge, MA, Harvard University Press, 1984.
6. C. Marchetti, On the Role of Science in Post-Industrial Society: 'Logos—the Empire Builders,' *Technol. Forecasting and Social Change*, Vol. 24, 1983, p. 197.
7. ————, *Innovation, Industry and Economy—A Top-Down Analysis* (publ. PP-83-006), Laxenburg, IIASA, 1983: Invention et innovation: les cycles revisités, *Futuribles*, March 1982; *A Post-Mortem Technology Assessment of the Spinning Wheel: The Last Thousand Years* (publ. PP-77-010), Laxenburg, IIASA, 1977.

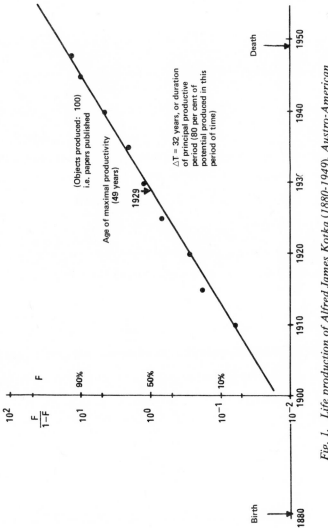

Fig. 1. Life production of Alfred James Kotka (1880-1949), Austro-American founder of mathematical demography. Data source: Elements of Mathematical Biology, New York, Dover 1956. This and the thirteen tables following were prepared by Cesare Marchetti, IIASA.

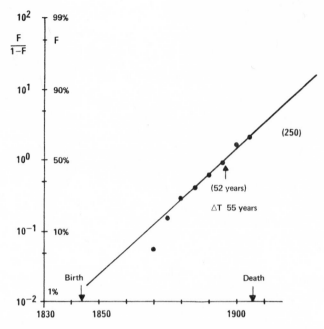

Fig. 2. *Production of Ludwig Boltzmann (1844-1906), Austrian physicist who introduced probability, linked to entropy, in thermodynamic processes.*

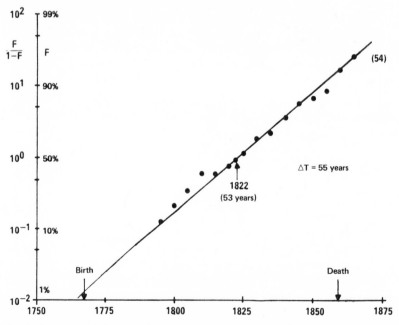

Fig. 3. *Production of Alexander von Humboldt (1769-1859), explorer and one of the founders of geobotany, climatology and oceanography.*

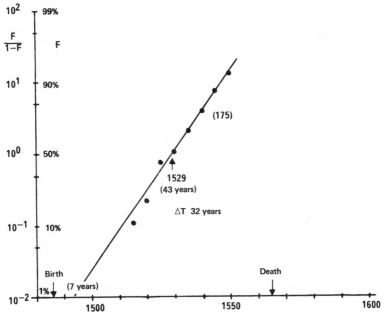

Fig. 4. Domenico Beccafumi (1486-1565), Tuscan painter, sculptor, mosaicist and metal founder, and builder of Genoa's Palazzo Doria. Data source: Beccafumi, Milan, Rizzoli, 1969.
Nota: Larousse gives his year of death as 1551, in Siena.

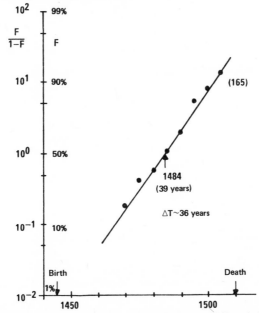

Fig. 5. Production of Botticelli (Sandro di Mariano Filipepi, 1445-1510), painter of mythological and poetic allegories for the Medicis. Data source: Botticelli, Milan, Rizzoli, 1969.
Nota: Larousse gives his year of birth as 1444.

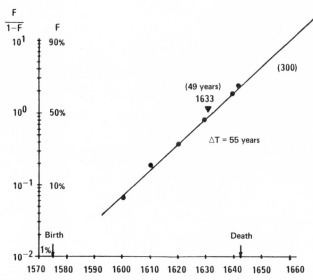

Fig. 6. Production of Guido Reni (1575-1642), sentimental and religious painter from Bologna. Data source: Guido Reni, Milan, Rizzoli, 1969.

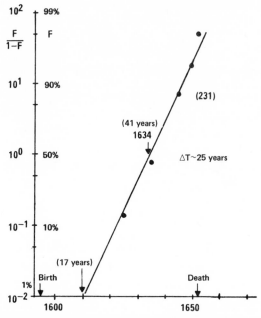

Fig. 7. Production of Jose de Ribera (1593-1652), Spanish painter who studied under Caravaggio (His Martyrdom of St. Bartholomew hangs in Madrid's Prado Museum.) Data source: Ribera, Milan, Rizzoli, 1969.
Nota: Larousse gives the birth date as 1588.

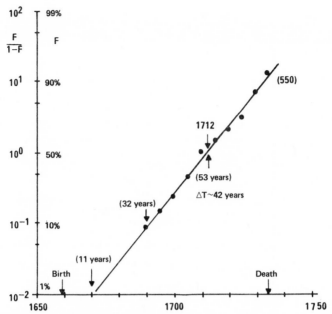

Fig. 8. Production of Sebastiano Ricci (1659-1734), an Italian baroque painter who did much of his work in Venice. Data source: S. Ricci, Milan, Rizzoli, 1969.

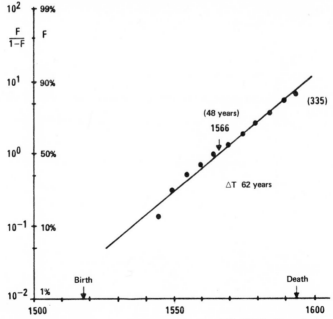

Fig. 9. Production of Tintoretto (Jacopo Robusti, 1518-1594), Venetian student of Titian.

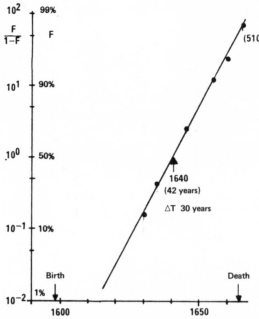

Fig. 10. Production of Francisco de Zurbaran (1595-1664), Spanish painter who
was influenced strongly by Carabaggio's style and by polychromatic sculpture.
Data source: Zurbaran, Milan, Rizzoli, 1969.
Nota: Larousse estimates the birth year as 1598.

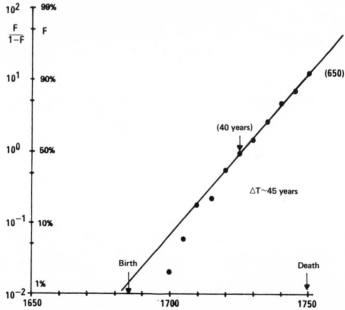

Fig. 11. Production of Johann Sebastian Bach (1685-1750), German violinist,
organist and composer. Data source: W. Schmieder, J.S. Bach, Wiesbaden, 1980.
The count is based on individual compositions and series, e.g., the Sechs Partiten
Klavierübung, counted as single items.

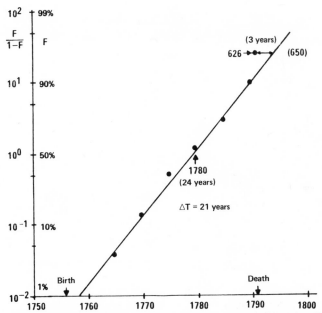

Fig. 12. Production of Wolfgang Amadeus Mozart, Austrian clavichordist, violinist and composer. Data source: L.R. von Köchel, W.A. Mozart, Wiesbaden, 1965.

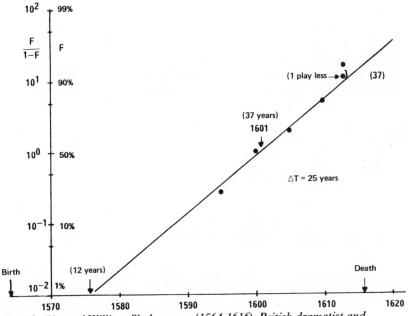

Fig. 13. Plays of William Shakespeare (1564-1616), British dramatist and dramatic poet. Data source: Enciclopedia Garzanti, 1982.

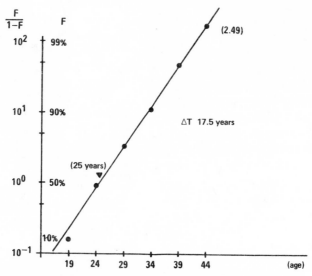

Fig. 14. The woman living in the United States produces an average of 2.49 children during her lifetime (80 percent during a productive period of 17.5 years; half the children are produced before 25 years). Source: Official statistics of the United States, showing cumulative fertility as of the year 1970.

Editor's note

Author Cesare Marchetti quotes a passage from the *Odes* of Horace. Because most readers will not have Dr. Marchetti's competence in Latin, the following rendition of the excerpt by the famous poet (65 B.C.-A.D. 8, approximately) should be of some help. The entire eleventh Ode is reproduced here, in translation.

Tu ne quaesieris

Do not inquire, we may not know, what end the gods will give, Leuconoe, do not attempt Babylonian calculations. The better course is to bear whatever will be, whether Jove allot more winters or this is the last which exhausts the Tuscan sea with pumice rocks opposed. Be wise, decant the wine, prune back your long-term hopes. Life ebbs as I speak: so seize each day, and grant the next no credit.

> *Horace, The Complete Odes and Epodes with the Centennial Hymn* (trans. W. G. Shepherd, with intro. by Betty Radice), Harmondsworth, Penguin Books, 1983, p. 79.

The two cases discussed in this chapter, the discovery of the double helix and the development of the theory of evolution through natural selection, support the incremental view of creativity. In both cases, the investigators were initially drawn to the problem because of the importance placed on it in the scientific environments in which they developed. Second, the early theorizing in both cases was a relatively direct extension of ideas available at the time. Because these initial theories were incorrect, a long series of revisions was begun as various aspects were brought into agreement with new information. Finally, the revisions and elaborations of the theories occurred in small steps rather than leaps of insight.

Chapter 5

The myth of scientific creativity

Robert Weisberg

The accompanying text appeared originally as Chapter 6 of Creativity, Genius and Other Myths *by Robert Weisberg; it is reproduced here with the kind permission of the book's copyright owners, W.H. Freeman and Company of New York. Professor Weisberg, who obtained the PhD from Princeton University in 1967, teaches psychology at Temple University, Philadelphia. He subtitled his volume,* What You, Mozart & Picasso Have in Common, *showing throughout its pages that much that we believe about the creative process is untrue. The entire work is highly recommended reading material.*
ISBN 0-7167-1769-7, available also in paperback. © 1986.

Great scientific discoveries have profound effects on every aspect of our lives, and the discoveries discussed in this chapter—Darwin's theory of evolution and the discovery of the structure of DNA—are clear examples of this. Darwin's theory of evolution through natural selection raised questions about the validity of religious beliefs important to Victorian society and about the position human beings held in the world, questions which arouse strong emotions even today. The discovery of the structure of the DNA molecule played a central role in a series of revolutionary advances in biology, one of which, the creation of new life forms, has wide-ranging implications for our lives. In these cases, and obviously in many others, the world has been drastically changed because of the creative work of scientists.

Perhaps because of this widespread influence of scientific theories, there is sometimes a tendency to take a very romantic view concerning the processes involved in scientific creation. Descriptions of scientists' creative thinking often emphasize the adventure involved, i.e., the theorist venturing alone into unknown territory, relying on nothing more than scientific intuition. Because the results are so extraordinary, it seems reasonable to assume that the thought processes are extraordinary as well.

The following is a description by James Adams of the work of James Watson and Francis Crick, the discoverers of the structure of DNA.

> Watson and . . . Crick relied heavily on inspiration, iteration, and visualization. Even though they were superb biochemists, they had no precedent from which they could logically derive their structure and therefore relied heavily on left-handed [intuitive, nonlogical] thinking.

Adams emphasizes the fact that Watson and Crick could not rely on past work, and so were forced to use their intuition to create something completely new. According to Adams, Watson and Crick apparently plunged into the unknown on their own without guidance from earlier scientific work.

Other writers also have told of creative scientists working independently of what was already known. Indeed, great importance is sometimes placed on the fact that such men and women lacked formal education, because formal education is assumed to stifle the creative processes. Lack of education presumably leaves the scientists' thought processes more 'free,' since they do not become wed to the old way of thinking about things. An example of this view is seen in the following quote, from Edward De Bono.

Many great discoverers like Faraday had no formal education at all, and others, like Darwin or Clerk Maxwell, had insufficient to curb their originality. It is tempting to suppose that a capable mind that is unaware of the old approach has a good chance of evolving a new one.

This chapter subjects this view of scientific creativity to critical analysis by examining two case histories of scientific discovery in detail. I shall try to show that scientists do not make great intuitive leaps into the unknown independently of what has come before, but that even in its most impressive manifestations, scientific discovery develops incrementally and is firmly based on the past. The discussion addresses the following sorts of questions. Why was the scientist interested in the phenomenon in the first place? What ideas concerning the phenomenon were already available? What was the relation between the new discovery and the old ideas? Did the new discovery unfold all at once, or as a series of steps? How did the discovery develop; did it grow from one single line of thought or were there backtrackings and changes of direction and, if there were, what brought these changes about?

I hope to show that scientific thinking progresses in a manner [whereby the problem investigators] first attempted to solve the problems directly, based on their knowledge of the problem situation or situations like it. Creative solutions developed as the problem solvers acquired information indicating that their initial solutions were inadequate. In attempting to overcome these inadequacies, subjects were led to try things they had not tried before.

For the present discussion, the important points are that the initial attempt to solve a problem depends relatively directly on what the person knows about the problem when he or she starts working. Changes in the way the person approaches the problem (that is, 'restructurings') occur in response to information that becomes available as the person works on the problem. That is, restructurings are not intuitive leaps into the unknown, but responses to changes in the problem. Finally, novel solutions to problems also arise in response to information that becomes available as the person works on the problem.

In addition, it is sometimes argued that scientific genius is somehow blessed with an intuitive sense of just which scientific problems are both potentially important *and* solvable. According to this view, the nongenius either spends time solving problems that are not particularly important, or tries to solve those that turn out to be unsolvable or not easily solvable. By considering the backgrounds of creative scientists, I hope to show that their interest in the problem that made them famous evolved naturally from the social and intellectual environment in which they developed. Once again, no extraordinary processes are needed to explain why the scientists were interested in certain problems. The problems in question were 'in the air' at the time, and they were not the only scientists to see the potential importance of those problems.

Thus, this chapter attempts to demonstrate that the thought processes involved in two important scientific discoveries were not very different from those used by ordinary people dealing with small problems. Elsewhere, I present a similar discussion of creative thinking in the arts.

Restructuring in the Discovery of the Double Helix

One particularly clear example of restructuring in scientific discovery is seen in Watson and Crick's discovery of the structure of the DNA molecule. The double helix of DNA, shown in Figure 1, is found in every introductory text in genetics, psychology, biology, and many related fields. Watson and Crick's discovery helped bring about the recent revolutionary advances in molecular biology and genetics. The discovery of the double helix did not unfold smoothly in a single direction; however, there were a number of false starts and much revision of early ideas before things finally fell into place.

Watson, who was in his early twenties when this work was conducted, had already received a Ph.D. in genetics from the University of Indiana. Watson's professor, Salvador Luria, believed that understanding how the genes controlled heredity depended on understanding their chemical structure. Watson was therefore sent to the laboratory of Herman Kalckar, a Danish chemist. While in Europe, Watson attended a scientific meeting where he saw an X-ray of DNA in crystal form, taken by Maurice Wilkins, an X-ray crystallographer from King's College, London. X-ray crystallography is the 'photographing' of crystals using X-rays. The technique involves taking a crystal of some substance and exposing it to a concentrated X-ray beam. The X-rays are deflected by the atoms in the crystal and the pattern of deflection is recorded on film, since X-rays will expose film. The procedure is also called X-ray diffraction because the X-rays are diffracted by the crystal. When the exposed film is developed, an expert can tell from the pattern formed by the X-rays how the atoms in the crystal are organized.

Wilkins's picture excited Watson greatly, because it showed him that genes could crystallize, or form into crystals, like salt crystals. In order for this to occur, the molecules of the material must be able to combine in a uniform way. If so, genes must have a regular structure that can be analyzed reasonably directly. At about this time, Linus Pauling, the

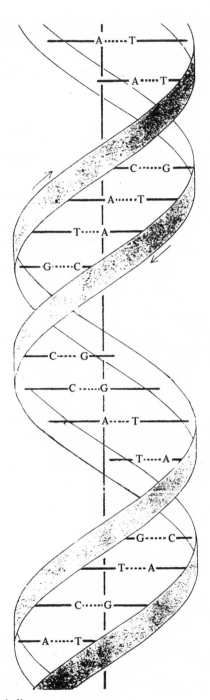

Fig. 1. The double helix.

internationally famous chemist, had proposed that the molecular structure of a protein was a long chain of atoms, which formed a helix (the shape of a spiral staircase). Pauling's work made Watson even more interested in learning how to work with X-ray diffraction in order to take pictures of crystals, so he obtained a position at the Cavendish Laboratory in Cambridge, England.

The Cavendish Laboratory was directed by Sir Lawrence Bragg, who had won the Nobel Prize thirty-five years earlier for his work in X-ray diffraction. Bragg and his colleagues had been applying this method in order to analyse larger and larger molecules. Crick had been on the staff at the Cavendish Laboratory for several years when Watson arrived, and was familiar with the use of X-ray diffraction, and considered it an important tool. A Ph.D. student in his mid-thirties, Crick had been trained in physics (as had Bragg) and was especially interested in developing theories concerning the structures of complex molecules. Thus, Pauling's work had impressed both Crick and Watson, and both had much interest in 'solving the problem' of the structure of DNA. Crick had also seen Wilkins's X-ray of DNA, and had talked to him about the molecule's possible structure.

Thus, when Watson and Crick met at Cambridge University in the fall of 1951, both of them believed that solving the problem of the structure of DNA was a task of great importance. This belief had been fostered by the scientific milieu in which each of them developed.

They then faced the question of where to begin. They took their general method from Pauling's work, which involved building models of molecules, and this pointed to the important evidence provided by X-ray photography concerning the structure of molecules. In addition to the general method of building molecular models, Pauling also provided some specific information about the kinds of structures that might be important. Since Pauling's helical model had already been successful, and because other possible structures would be much more complicated than a helix, Watson and Crick decided to try to model DNA using helical models. They set out to build a model of the DNA molecule that was consistent with all that was known about DNA at the time. This included information from studies of its chemical composition, the sunburst patterns of X-ray photographs which revealed bits and pieces about the shape of the molecule, and evidence from genetic studies on how DNA worked during reproduction.

Given their backgrounds and knowledge, it is not hard to understand why Watson and Crick chose to work on DNA in the way they did. These facts raise questions about the Adams claim that Watson and Crick had 'no precedent' to work from. Pauling and Bragg provided much in the way of methods and Pauling provided specific starting points concerning possible structures. As will be shown, still other information was provided by other scientists.

These preliminary decisions led to several further questions for Watson and Crick. First, how many strands should the helix contain? There was evidence from X-ray photographs that the molecule was thicker than a single strand, but did it contain, two, three, or four strands (Figure 2)? There also was the question of whether the backbones of the strands were located on the inside or outside of the molecule (Figure 2). In a spiral staircase, the backbone is located on the outside and the steps of the staircase are located on the inside. The steps of the staircase can be considered equivalent to the *bases* of the DNA molecule, which actually carry the genetic message. It was also possible, however, that the backbones were located inside the molecule and the bases on the outside.

Near the end of November, 1951, Watson and Crick built their first model, a three-stranded helix with bases on the outside. This model was based on various pieces of information that were available about DNA, perhaps most importantly on work being conducted by Wilkins's group at King's College. These two teams were more-or-less friendly rivals. Watson recently had attended a lecture by Rosalind Franklin, one of Wilkins's colleagues, in which she discussed X-ray pictures of DNA and the amount of water present in the molecule. Watson and Crick discussed Franklin's results (or, rather, they had discussed Watson's memory of them) and decided to start with a three-strand, center-backboned model because they felt that only such a structure would be regular enough to produce the clear X-ray pictures that the King's group had obtained. The specific shape of the helix, with the three backbones twisting about each other, was designed to enable it, among other things, to hold the amount of water Franklin reported.

Soon after this initial model was completed, the King's group visited Cambridge to see it. The meeting was a disaster for Watson and Crick. First and foremost, Watson misrecalled Franklin's report of the amount of water in DNA, so the model contained only one-tenth of the necessary water. This and other problems made it obvious that this three-stranded model was simply incorrect.

To summarize this initial phase of Watson and Crick's work, they adapted Pauling's method and attempted to apply it to a similar type of problem. The initial solution was inadequate and they were forced to modify it.

Over the next fifteen months or so, basic changes took place in Watson and Crick's thinking and model building. The three-strand,

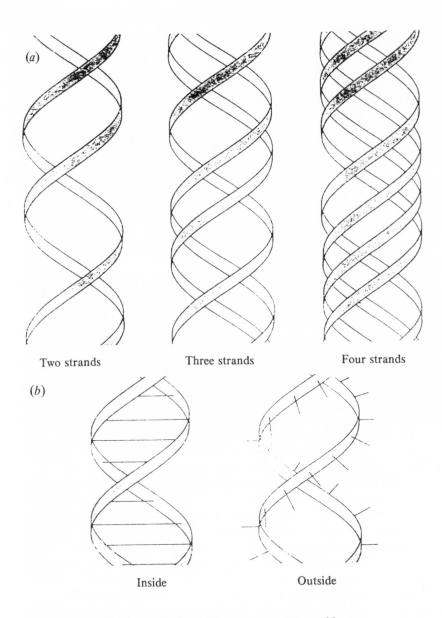

Fig. 2. (a) Multiple-strand helices. (b) Possible positions of bases.

center-backbone model became a two-strand, outside-backbone model. Thus, two restructurings took place—in the number of strands, and the location of the backbones. How did they come about?

As far as the position of the backbones was concerned, several further pieces of evidence pointed to an outside-backbone structure. First, late in 1951, as Watson and Crick tried to revamp their three-strand inside-backbone model to deal with the problems raised by the King's group, all the inside-backbone models they devised violated basic laws of chemistry. In order to build an inside-backbone model, atoms had to be put closer together than the laws of chemistry allowed. This problem probably impressed Watson and Crick greatly, because much of Pauling's work had been spent accurately measuring the distances between atoms in molecules. With their knowledge of Pauling's work, therefore, Watson and Crick would very likely have worried about distances between atoms. Second, early in 1952, Wilkins wrote to Crick saying that he was almost certain the backbones were on the outside. Third, in January, 1953, Watson saw a new X-ray photograph of DNA made by Franklin and, to the expert eye, the backbones were clearly on the outside. (Watson learned to read X-ray photographs in 1952.)

During this time, other factors began to point more and more strongly to a two-stranded structure rather than three strands. First, several other investigators working on three-strand models had also failed to produce anything. Indeed, early in 1953 Pauling himself proposed a three-strand, center-backbone model, much like Watson and Crick's, and it turned out to be just as wrong. Second, though the two-strand structures would not contain as much water as Franklin's measurements indicated, it was possible that her measurements were incorrect, which would make two-strand structures more plausible. Third, information in a report by Franklin indicated that the backbone chains came in pairs and ran opposite to each other, leading one to expect two chains rather than three. Of course, it was possible that there were four, six, or eight strands, etc., but if there were an even number of strands, two seemed the obvious place to begin model building. Finally, one reason why Watson had not wanted to try outside-backbone models was because he had not been able to see how to fit the bases inside so that the structure would be regular enough. As a rough analogy, Watson was trying to build a spiral staircase, but he was not sure that all his steps were the same size. With different-sized steps, the helix would be irregular, would wobble, and would not produce clear X-ray pictures. Early in 1953, however, the models of the bases were not yet available from the machine shop in the Cavendish Lab, and Watson could ignore the bases and try to construct a two-strand, outside-backbone helix that would fit the evidence available.

To summarize the work so far, the initial model was abandoned because of various problems with it. These problems, as well as several new pieces of data, led Watson and Crick to begin work on a different type of model. The two crucial restructurings took place in response to changes in the problem situation.

Watson had little trouble constructing a two-stranded helix without bases, which, if DNA were similarly constructed, would account for Franklin's X-ray photographs. The next question involved the bases—a way had to be found to fit them together to form the steps that held the backbones together. Once again a restructuring had to take place, and once again the restructuring occurred in a straightforward way.

Watson first tried putting the bases into the center of the model one way, but this turned out to be incorrect. He had tried to pair bases of the same type, called 'like-like' pairing, to make the steps which would hold the backbones together. The problem was that the various bases were of different sizes, so that when put together, the steps were also of different sizes, and the backbones could not be the same distance apart at all points. There were additional reasons for rejecting the like-like method, so Watson then tried various combinations of different pairs of bases, essentially by trial and error, until he found the needed combinations. Here is how Watson describes the final steps.

When I got to our still empty office the following morning, I quickly cleared away the papers from my desk top so that I would have a large, flat surface on which to form pairs of bases held together by hydrogen bonds. Though I initially went back to my like-like prejudices, I saw all too well that they led nowhere. When Jerry came in I looked up, saw that it was not Francis, and began shifting the bases in and out of various other pairing possibilities. Suddenly I became aware that an adenine-thymine pair held together by two hydrogen bonds was identical in shape to a guanine-cytosine pair held together by at least two hydrogen bonds. All the hydrogen bonds seemed to form naturally; no fudging was required to make the two types of base pairs identical in shape . . . . Upon his arrival Francis did not get more than halfway through the door before I let loose that the answer to everything was in our hands.

While fiddling with the base models, Watson saw that two specific pairings produced steps of identical size and would thereby serve to hold the two backbones a constant distance apart. Furthermore, these particular pairings turned out to match the results of earlier studies of the chemical composition of DNA, which was further reason to believe that the pairings were correct. Also, the structure made it easy to understand how DNA replicated itself during cell division, which was still another point in its favor.

In summary, after more than a year of work and thought, the final step in the discovery came about very quietly. The earlier ideas on pairings proved inadequate, which led to another, relatively limited set of possibilities. Watson tried these possibilities until he hit the one that worked.

One obvious aspect of this discovery is how it differs from the romantic view of scientific creativity considered earlier. There were no soaring leaps of insight, no sudden awakenings in the middle of the night with the long-sought solution, no mysterious appearances of ideas from out of the mists of the unconscious. If Watson's account is reasonably accurate (and Horace Judson supports Watson in the points discussed in the last few sections), the sequence of events is very ordinary, much like what happens when an average person tries to solve a simple problem. Based on the backgrounds of the participants, various possible solutions were considered, and modified, until the final solution was produced. As in the various laboratory situations discussed earlier, one can see how the solution evolved in a series of conscious steps, as Watson and Crick worked on the problem. One can also see how the problem changed as they worked on it, as they acquired more information about DNA and learned of the inadequacies of their earlier ideas. In addition, as they got closer to the solution, there were no new methods in their work, nor any striking changes in their results. They went forward a little bit at a time, and sometimes they went backward.

Although this is one of the great discoveries in modern science, the methods are surprisingly ordinary. The *result* is what makes the whole episode extraordinary, not the methods. As one looks at the final product, the DNA model, however, it is sometimes difficult to understand how such a conception could ever be produced by ordinary thought processes. With Watson's report as a guide, however, things become much clearer. Interestingly, Watson reports that when he was struggling with the base pairings, he went to the movies in the afternoons, hoping that the answer would suddenly appear in an illumination as a result of incubation. Nothing came, however, nor did he enjoy the films much because he had trouble forgetting about his work.

It should be emphasized that though one claims no extraordinary thought processes were involved in the discovery of DNA, not just anyone could have produced that discovery. The particular scientists and circumstances involved were unique, and all contributed to the discovery. First, both Watson and Crick were committed to Pauling's model-building method which, although things did not always unfold smoothly, turned out to be very useful. There is some evidence that Rosalind Franklin was also very close to discovering the correct structure of DNA, but one reason that she did not get there when Watson and Crick did may have been that she was less committed to model building. Second, both men felt that solving the DNA puzzle would be of great scientific importance and probably make them scientific immortals—this made the whole undertaking exciting and kept them working at it. Third, coming from different scientific backgrounds enabled them to criticize each other's ideas from a different perspective. Other people at the laboratory in Cambridge also added their expertise at critical times. Finally, both Watson and Crick seemed willing to take risks in their thinking, and while this sometimes got them into trouble, it also ultimately led them in the correct direction.

Thus, when one analyzes the factors that contributed to the discovery of the structure of DNA, one sees that the specific people involved were just one of several factors of critical importance. *These* scientists, working on *this* problem, in *this* setting, produced the discovery. If any of these factors had been significantly changed, someone other than Watson and Crick probably would have made the discovery. This leads to the conclusion that scientific creativity is much more complicated than an isolated genius working on a problem, which supports the analysis of scientific genius presented elsewhere. Such complexity is seen even more clearly in the case of Charles Darwin's discovery of the theory of evolution through natural selection. Although Darwin worked essentially in isolation for the fifteen months during which he thought intensively about evolutionary theory, his work also was strongly influenced by that of others.

Darwin and natural selection

The theory of evolution through natural selection is sometimes presented as an example of a great scientific discovery occurring in a leap of insight. Indeed, Charles Darwin's *Autobiography* seems to support this view, especially in the following well-known passage in which he discusses how Thomas Malthus's *Essay on Population* influenced his thought. In this essay, Malthus presents the view that any population grows faster than its food supply, resulting in many members of the population being unable to find food. Darwin comtemplated Malthus's argument and realized that it meant that some animals which were 'more fit' would survive in this competition for food. This in turn would result in these animals passing on their characteristics to their offspring, and so the population would evolve.

In October 1838, that is, fifteen months after I had begun my systematic enquiry, I happened to read for amusement Malthus on *Population*, and being well prepared to appreciate the struggle for existence which everywhere goes on from long and continued observation of the habits of animals and plants, it at once struck me that under these circumstances favourable variations would tend to be preserved and unfavourable ones to be destroyed. The result of this would be the formation of a new species. Here, then I had at last got a theory by which to work.

Darwin's reading of Malthus allowed him to formulate in essentially complete form the theory of evolution through natural selection. Soon after reading Malthus, Darwin wrote the following in his notebook.

Three principles will account for all 1. grandchildren like grandfathers, 2. tendency to small change especially with physical change, and 3. great fertility in proportion to support of parents.

This is the encapsulation of the theory of evolution by natural selection. First, each generation passes its characteristics to the following generations. Second, organisms within a given generation vary in many different small ways. Third, because of parental fertility there are many more offspring than parents. This means that offspring must compete for limited resources, and that any variation that helps a given organism compete and survive will be passed on to the next generation. Thus, species constantly evolve due to the 'selection' produced by competition.

The passage in Darwin's autobiography, in conjunction with the evidence from the notebook, seems to describe a creative leap to a new theoretical formulation. However, other evidence indicates that the incident was not the leap it appears to be. My description of how the theory of natural selection evolved in Darwin's thinking, hopefully, will show how his views changed over the years as he thought about evolution. Once again, the discussion is concerned with such questions as how Darwin's interest in the question of evolution developed, what ideas were already available, how Darwin's ideas were related to older ideas, and what steps were involved in his formulation of his theory. As occurred with the discovery of the double helix, Darwin did not make steady progress toward his final theory—his early views had to be partly abandoned and partly modified before progress could be made. In addition, Darwin was influenced greatly by the thinking of others, regarding both his general interest in biology and the question of evolution and his more specific views concerning how evolution came about. Before presenting Darwin's discovery, it is first necessary to place it in a historic and social context.

Interest in the possible evolution of species did not originate with Darwin. For several centuries, scientists and philosophers in Europe had been contemplating questions of how species originated and changed over time. As one example of evolutionary theorizing, the work of the Comte de Buffon (1707-1788) is particularly interesting because it probably mentions every significant ingredient in Darwin's theory. Buffon's *Histoire Naturelle*, 1749, discusses the following important factors. First, life sometimes multiplies faster than the food supply, thus producing a struggle for existence. Second, there are variations in form within a single species (no two organisms are identical). Furthermore, these variations are often inheritable and can be taken advantage of by carefully breeding stock. (*Artificial* selection, the human control of animal and plant development through selective breeding, was well-known in Europe long before Darwin.) Third, there is an underlying similarity of structure among animals that are very different, which hints at evolution from a common ancestor. Fourth, long stretches of time are necessary to explain how life on earth developed. Fifth, some animal life has become extinct. Finally, Buffon's overall philosophy is oriented toward an experimental approach to the study of questions concerning evolution.

Buffon is only one example, though perhaps the most striking, of interest in the questions that were to concern Darwin. Furthermore, the tentative answers he proposed were similar to those ultimately proposed by Darwin. Thus, European scientific circles had already developed a deep interest in evolution, which means that Darwin's interest in these questions is by no means extraordinary. Furthermore, Darwin had an even stronger personal reason for interest in evolution due to his family.

Darwin was born in 1809 and died in 1882. During his long life he published many articles and books in addition to *The Origin of Species*. Though he died before Charles was born, his grandfather, Erasmus Darwin, developed a theory of evolution based on the inheritance of acquired characteristics. The elder Darwin proposed that during their struggle for survival some animals developed characteristics that enable them to adapt better to the environment. These characteristics were handed down to the next generation and evolution thereby occurred. (A very similar view was independently proposed by Jean Baptiste Lamarck, and this general viewpoint is called Lamarckian evolution.) Thus, the problem of evolution was familiar and important to young Darwin, since not only had his grandfather written about it, but the members of his family probably discussed it. In this respect the Darwins were similar to a significant number of people with liberal views who were interested in scientific answers to questions involving the origins and evolution of life on earth.

It is important to contrast the evolutionary view with the prevailing view of the time which relied on the Bible to explain the origins of life on earth. According to the strictest version of the biblical view, all life was brought forth when God created the heavens, the earth, and all living things. One aspect of this literal view of the Bible was that each of the various species was a perfect creation and therefore not subject to change, making any discussion of evolution blasphemous as well as unnecessary. When discoveries of fossils and rock strata raised questions about God's creation, Noah's flood was made to account for them. When it was argued that the organization of multiple fossil layers made it unlikely that they could have been laid down during such a flood, the biblical view was modified to include many floods, of which Noah's was simply the latest. Thus, the orthodox view was that there was no evolution, species were static and unchanging, and periodic catastrophies wiped out all life.

In scientific circles there were many who did not accept this orthodoxy, and Darwin met many such thinkers during his studies at the universities of Edinburgh and Cambridge. He entered Edinburgh University in 1825, to study medicine, but became upset by various aspects of his education and left the university in 1827. While at Edinburgh, he met and became close friends with Dr. Robert Grant, a zoologist who believed in Lamarckian evolution. Darwin and Grant spent much time together and talked about Grant's views on evolution.

Darwin attended Cambridge University from 1827 to 1831. Here, too, he met and was influenced by many important scientists. One of these, John Stevens Henslow, professor of botany and geology, spent much time talking with Darwin and also welcomed him to the weekly open house he held that enabled students and professors to get to know each other and share ideas. Professor Adam Sedgwick invited Darwin on a geologic expedition to north Wales in August, 1831. The scientific techniques of observation and data collection Darwin learned on this trip were important for the rest of his scientific life.

In summary, although Darwin says in his autobiography that he did not get much out of his classes at Cambridge, he seems to have gotten a full education and, by the time he left, was no longer a young man ignorant of modern science. He was able to embark on the next great adventure of his life, the voyage of the *Beagle*, as a scientist sophisticated both in methods of data collection and in modern scientific theory, including modern evolutionary views.

On August 29, 1831, Darwin was offered the post of naturalist on HMS *Beagle*, which was scheduled to carry out a five-year journey around the world, paying particular attention to exploring the shores of South America. The post of naturalist involved collecting and cataloging specimens of animal and plant life. The *Beagle* sailed from England on December 27, 1831, and returned October 2, 1836. During the voyage,

Darwin acquired information that convinced him that evolution was a fact, and which set the stage for his attempt to solve the problem of how evolution occurred.

There were two sorts of influences on Darwin during the voyage. The first came from his reading of Charles Lyell's *The Principle of Geology*, and the second came from some of his observations, most particularly those of the animals inhabiting the Galapagos Islands. Lyell's book presented a strong case against the view that science should be based on a literal interpretation of the Bible. Darwin was in the perfect position to gather data concerning phenomena discussed by Lyell in support of the idea that species did not stay the same but continuously evolved. Lyell talked about the competition among plants and animals, and how unhealthy organisms would be destroyed by healthier ones in the 'struggle for existence,' a phrase that seems to have been used first by Lyell.

Interestingly, Lyell used natural selection as a principle of evolution, but only in a negative manner, as when unhealthy animals lose the struggle. It remained for Darwin to realize that natural selection also worked in a positive manner, since any change that gave an animal an advantage in the struggle for existence would be passed along and thereby result in the evolution of the species. One reason Lyell failed to see this may have been that he was unaware of the tremendous variation that could occur within a species. Darwin became aware of this potential for variation during the voyage of the *Beagle* and therefore was in a position to expand upon Lyell's theory.

As his diary and notebooks show, Darwin was leaning toward a belief in evolution when he left on the voyage, and this idea was strengthened by his observations during the voyage. As the *Beagle* traveled south along the coast of South America, for example, Darwin noted that a series of similar animal groups occurred, with the replacement of one group by another of very similar form. Such a series of similar groups suggested that a single species had differentiated into several highly similar forms. In addition, while exploring the pampas of Argentina he discovered fossils of huge animals that were anatomically very similar to the armadillos existing at that time in the region. This supported the notion that the modern species gradually evolved out of the ancient species. These similarities, which were based on space and time, were somewhat surprising if one believed the literal theologic view that all species were created at once. Why had the creator made these various species so similar? If one believed that these species were related to one another, then these similarities made some sense.

When the *Beagle* reached the Galapagos Islands, a group of approximately twenty islands off the west coast of South America, in

September, 1835, Darwin found even more remarkable similarities and differences among species of animals. He collected animals from the various islands and found that the local inhabitants could tell which island each specimen had come from. Thus, the species of animals on these islands differed from each other and were not stable. Darwin was particularly impressed by the various species of finches on the islands, especially the variations among their beaks. Some species had small beaks, while others had large thick beaks, differences which pointed to a great variability within groups of animals, even when the physical environment was essentially identical as it was on these islands. Thus, Darwin was faced with the question of how, by what mechanism, all this variation came about.

In summary, the voyage of the *Beagle* pushed Darwin's thoughts toward the theory of natural selection in several ways. He came to believe that evolution did occur, and he saw that species could vary significantly in their characteristics, even within a constant environment. It is important to keep in mind that during the five-year voyage of the *Beagle*, questions concerning evolution of species were central in Darwin's mind. This provided ample opportunity for his ideas to be changed.

Darwin began two important tasks after his return to England in October, 1836—organizing the material collected on the voyage and thinking systematically about evolutionary theory. In July of the following year he began his first of four notebooks on transmutation of species (evolution), and he formulated his first theory of evolution based on the idea of the 'monad.' Though this theory is relatively far removed from the theory of natural selection, it is possible to see how and why Darwin changed it until over a year later he arrived at his final theory.

The monad theory is based on the idea that simple living particles, or monads, are constantly springing into life. Monads originate in inanimate matter and are produced by natural forces, so that one does not have to assume that there is a separate supernatural creation for each monad. Each monad has a fixed life span during which it differentiates, matures, reproduces, and dies, becoming a whole group of related species. The simple initial particle becomes more complex over time and ultimately forms the complex organisms living today and preserved as fossils. These organisms respond to changes in the environment with adaptive changes. Thus, changes in the environment produce changes in the monad's 'offspring,' or evolution. These organism-environment encounters are essentially random, so monad development is irregular. Since not everything that could happen does happen, some evolutionary possibilities do not occur. When a monad dies, all the species it has become die at that same moment. The total number of species stays approximately constant, since when a monad and all its related species dies, new species will develop to replace them.

This theory has components that seem very primitive today, but were accepted at the time. The idea of the spontaneous generation of life from nonliving matter, for example, was not disproved until Pasteur's experiments in 1861. Lamarck also argued that spontaneous generation occurred continuously, while Lyell claimed that species were created in succession and endured for a fixed period. Nor was the monad concept new with Darwin. He took these beliefs and used them as the basis for his formulation of how extinction of species might occur.

Furthermore, at that time, there were several different reasons for believing that extinction occurred through the simultaneous deaths of all the products of a single monad. First, fossil evidence indicated that whole groups of species disappeared suddenly. This 'evidence' was actually a mistaken conclusion drawn from imperfections of the geologic record, which in reality is very fragmentary. At the time, however, Darwin accepted it as valid. Second, in order for extinction to come about through *environmental* change, which would make the belief in monads unnecessary, a relatively large-scale change in the environment was needed. This view, a version of catastrophism, which maintained the great upheavals occurred periodically and changed the world drastically, had been rejected by Darwin. Also, if environmental change rather than monad death produced extinction, any change in a species could be erased by a change in the environment. This would mean that there would be no overall evolution. Darwin, therefore, looked for a nonenvironmental cause for extinction (monad life span).

Thus, Darwin's monad theory was a reasonable attempt to deal with the facts of species change as he knew them, and it is firmly based on other accepted ideas. Over the next fifteen months or so, great changes occurred in Darwin's thinking, that is the monad theory gradually changed into the theory of natural selection.

Before Darwin was pushed to the theory of natural selection by reading Malthus, a number of changes in his views had resulted in his abandoning the monad theory. These changes came about through his logical analysis of the consequences of his beliefs, new sources of evidence, and a change in emphasis concerning what his theory should account for.

One important aspect of the monad theory was its attempt to account for the origin of life through the development of monads. Darwin's interest shifted to the possibility of life as an ongoing system, however, rather than a system involving the constant creation of new forms. One important factor in this change was the discovery of fossils of unicellular organisms, which meant that some organisms remained simple and that the development of organisms from simple to complex, as assumed by the monad theory, did not always occur. One did not have to assume that all simple organisms alive at present had just been created.

In Darwin's early thinking, the notion of variation among the members of a species was a conclusion drawn from the fact that changes in the environment produce changes in organisms, that monads develop into complex organisms, and that accidental encounters of the organism and the environment result in new evolutionary lines. In Darwin's final theory, variation is accepted as a premise ('tendency to small change . . .'), indicating a basic shift in viewpoint.

There seem to be several reasons for Darwin's shift in opinion concerning the role of variation in his theory. First, a theory of evolution that assumes an inherent tendency for organisms to become more complex and adapt to changes in the environment may say nothing more than evolution occurs simply because it occurs. That is, these assumptions may not address all the important questions of how evolution actually comes about. Also, Darwin became increasingly impressed by the extent of variation in nature, much of which did not seem to be in direct response to environmental change. His experiences on the *Beagle*, especially in the Galapagos, showed him that animals in the same environments can differ, indicating that such variations are not in response to environmental changes. Variation is a given, a fact of organic life.

One thing that became increasingly important in Darwin's thinking is the great fertility (fecundity) in nature, the fact that organisms reproduce at a very high rate. This *super*fecundity was emphasized by Malthus in his *Essay on Population*, but was referred to by other sources as well. The German biologist C.G. Ehrenberg, who was becoming well known in British scientific circles at the time, provided evidence that micro-organisms reproduced at nearly-unbelievable rates. As an example, Darwin wrote in his notebook shortly before reading Malthus that a single micro-organism could produce enough offspring in four days to form a stone of considerable size. Furthermore, the Malthusian idea of large numbers of offspring was discussed by many authors with whom Darwin was familiar, such as Erasmus Darwin and Lyell.

It is important to emphasize again that the idea of natural selection did not originate with Darwin. It was generally acknowledged that deviant organisms tended to be less fit and therefore had less chance of survival. The idea of natural selection as 'nature's broom' sweeping away weak organisms had been discussed by many important theorists before Darwin. Darwin's great contribution was the realization that natural selection could also work in a positive direction. In addition, Darwin was familiar with the phenomenon of artificial selection, in which humans breed plants and animals for special characteristics. Though an analogy can be made between the human role in artificial selection and the struggle for survival in natural selection, when Darwin began theorizing about evolution this analogy was not clear to him.

Rather than being a great leap of insight, reading Malthus was simply the final step in a long process during which Darwin's views underwent many changes. These changes in thinking were made up of small modifications based on pressure from new data or logical problems which became apparent to him. Furthermore, the Malthusian insight itself did not originate from nothing, that is, Darwin's views had to change before Malthus could have just the right effect on him.

There is also some evidence that Darwin did not originally perceive his response to Malthus to be a great leap of insight or a profound unveiling of something completely hidden. In the passage from Darwin's autobiography presented earlier, Darwin wrote 'it at once struck me . . . ,' which leads one to believe that everything fell into place at once. That passage, however, was written many years after the event actually occurred. The entries made in his notebooks fail to indicate that anything particularly momentous occurred. The entry that refers to his reading of Malthus *looks* no different from other entries, and other entries in the notebook made about that time continue to refer to other topics, indicating that Malthus's essay had not produced an illumination that had captured all his other interests. Finally, and perhaps most important, more than a month elapsed before Darwin wrote the 'Three principles will account for all' passage cited earlier, in which he succinctly summarized his theory. In other words, there still seemed to be a lot of working out to be done after he read Malthus, and one does not get the feeling that everything suddenly fell into place. Rather, Malthus was just another source of information and ideas which Darwin used as the basis for his thinking.

In conclusion, it seems that when Darwin wrote his autobiography, he may have forgotten exactly what occurred years before. This autobiography also contains another inaccuracy about his thinking. In a passage shortly before the one describing his Malthusian insight, Darwin says that when he started writing his notebooks, he worked without a theory and simply collected facts. The notebooks themselves show this statement to be false because Darwin produced the monad theory very soon after beginning the notebooks. It seems that Darwin wrote his autobiography without consulting his own notebooks.

To summarize, the evidence supports the idea that Darwin's thinking proceeded in a similar manner to that of Watson and Crick. His interest in the problem of evolution was not at all extraordinary; he was acquainted with the thinking of many others in this field, and their ideas were incorporated into his initial monad theory of evolution. The inadequacies of this initial theory required much modification before the theory of natural selection was produced. This modification came about as Darwin's views changed, as he thought more about the consequences of the monad theory, and as he acquired new information about the various phenomena involved. The important thing to emphasize is that there seems to be no need to postulate any extraordinary thought processes in order to understand Darwin's accomplishments. Though he was considering issues of extraordinary importance, he did so in very ordinary ways.

Conclusions

In at least two well-known cases, scientific creativity came about through the same sort of thought processes ordinary people use to solve ordinary problems. In both the discovery of the double helix and the development of the theory of evolution a series of similar phenomena occurred. First, the early theorizing, based relatively directly on ideas of others, was shown to be inadequate in several ways. These inadequacies were addressed by modifying various parts of the old theory, and the result was a new theory. These modifications did not occur all at once in a leap of insight, but rather involved a series of small accommodations as each difficulty was handled. There seemed to be no one point at which everything suddenly fell into place, but rather, there was a gradual closing in on the final theory as bits and pieces fit together. On the whole, the process discussed in this chapter is remarkably similar to the process of creative problem solving.

The conclusions drawn here are also relevant to discussion about the importance of divergent thinking in creativity, where it is concluded that creative problem solving does not involve suddenly shifting to a new way of viewing the problem. In neither case discussed in this chapter did anything like divergent thinking occur. The theorists involved began with ideas that were readily available in the scientific community, and they modified them to make them relevant to the specific problems they were trying to solve. When the initial formulations were shown to be inadequate, the new directions their thoughts took were not due to anything like divergent thinking, but were the result of the analysis of the new difficulties and the use of additional knowledge and logical reasoning abilities. Nothing like brainstorming seems to have occurred in the discovery of the double helix or the theory of evolution.

Additional studies of scientists also indicate that divergent thinking is not important in scientific creativity. In these studies, scientists of different degrees of creativity were given tests of divergent thinking. The consistent finding in these studies is that performance on divergent thinking tests is unrelated to scientific creativity. That is, the more creative scientists do not perform better on tasks involving divergent thinking, and scientists who do best on the divergent thinking tests are not the most creative in their professions. Thus, divergent thinking does not measure the abilities involved in scientific creativity.

The conclusion developed in this chapter—that creative thinking in science progresses through a series of incremental steps, perhaps in irregular directions—contradicts the generally accepted belief about scientists' thinking process, based on their own reports. Many scientists and creative thinkers in other areas have reported that their creative thinking is often brought about through leaps of insight. This leaves us with a problem regarding these subjective reports; the present analysis does not seem to apply to them. It may be simply that creative thinking occurs in more than one way, the small-step, incremental, revisionist method seen in this chapter, and the insightful-leap method. If this is true, however, it seems surprising that neither of two highly creative scientific discoveries involved such leaps. Furthermore, these two discoveries are extensively documented, making it reasonably clear that neither involved leaps of insight. Also, subjective reports of leaps of insight are often very inaccurate sources of evidence. Thus, whenever extensive documentation is available, such as an individual's notebooks, the evidence tends not to support the subjective reports.

It may be that the difficulty lies with the subjective reports of creative persons and, therefore, the analysis in this chapter does apply to the thought processes in all acts of creativity, scientific and otherwise. Subjective reports are usually difficult to verify because there is seldom any way of knowing if the person is correct or not. There is also evidence of subjective reports being deliberately falsified. Finally, experimental evidence shows that subjective reports can be mistaken. Darwin's report in his autobiography concerning his Malthusian 'insight' is a perfect example of a scientist's mistake in a subjective report. When a scientist tries to reconstruct from memory thought processes of a complicated nature that involved emotionally arousing work on an important problem, and which may have occurred many years before, there is a good chance that the reconstruction will be incorrect. If one relies on objective evidence, rather than on subjective reports, then there is no need to postulate leaps of insight. I conclude that the creative individuals who report great leaps of thought were simply mistaken. Since most of those individuals were neither in the business of studying their own thought processes (that is, they were not professional psychologists) nor concentrating on their thought processes at the time of creation, such mistakes are not surprising. ∎

PART II

Art, Music and Values

In mathematics, part of the Galois theory of mathematical groups finds special application in the demonstration of which algebraic equations are solvable by sequences of rational procedures and by the extraction of integral roots of previously known quantities — as well as in proving that a fifth-degree polynomial (or quintic) equation exists that cannot be solved by such operations. The theory can be applied to art forms via computer, as shown here by three Croatian specialists.

Chapter 6

Galois fields: anticipation, a different point of view, and some computer art

Vladimir Bonačić
and associates

Dr. Bonačić, Mr. M. Cimerman and Ms. D. Donassy can be reached at Postfach 24 01 01, 5300 Bonn 24 (Federal Republic of Germany), where their telephone number is 02244/4966. Dr. Bonačić states that the illustration and text 'illustrate why I have been preoccupied for twenty years with Galois theory.' He won the Nikola Tesla National Prize for Science (Zagreb, 1968), shared the Praemium Erasmianum (Amsterdam, 1975), won the Prize of the International Humanitarian Poster Exhibition (London-Baghdad, 1979), and is on the editorial board of Leonardo, *an art-and-science journal.*

Galois fields for the present

A focal point of computer art, as it is developing today, tends towards dematerialized and transcendental processes having timeless and spaceless qualities. The original computer graphic on which the accompanying illustration is based revealed an unknown, multi-perspective feature related to an abstract Galois field (named after the young French mathematical genius, Évariste Galois, 1811-32). The 'size' of computer graphics or the number of colours possible is practically unlimited. If a digital video processor is used, a time dimension is introduced and the same work of art is displayed as an animated computer-image using colours drawn from a palette of more than 16 million different shades.

In order to perceive a work of dematerialized computer art, time and space dimensions need to be added by the artist—using heuristic methods. The complexity of the problem can easily exceed the capacity of a single artist. Teamwork is thus advisable, and the impact of love (yes, love!) is essential for creation and inter-subjective communication among the team's members. The relationship between love, dematerialized computer art, inter-subjective communication (a well-defined phenomenon in existentialist philosophy) and common language—the future language for all of us and for artificial systems—is also of interest to those dealing with religious questions. (The Franciscan religious order, for example, has introduced computer art, computer music and a 'dynamic object' in their preparations for meditation at St. Kilian's Church, Wiesbaden, Federal Republic of Germany.)

The work shown here results from the collaborative efforts of Vladimir Bonačić (a cybernetician as well as artist), Miro Cimerman (a software designer), and Dunja Donassy (architect and city planner), who hail from Zagreb, Yugoslavia. Known commercially as *bcd*, the creators of the illustration shown are a team involved in designing computer imaging systems and computer imagery itself in the field of the socio-political sciences for use by the media (such as broadcasts during local and national elections, elections of the European Parliament), and in special television news and feature programmes. In applied economics, they develop thematic maps, using pattern recognition and colour topology to extract obscure or hidden data structures.

A little bit of background

The earliest known equivalent of algebraic equations is found in the so-called Rhind papyrus, evidently compiled from earlier works by the

Egyptian, Ahmes, sometime before 1700 B.C. Nothing really new transpired during the nearly four millennia since that time, only minor innovations within the same overall paradigm.

The theory of algebraic equations—a theory concerned with determining the solvability of an equation—dates only from the first third of the 19th century, a time when mathematicians could not solve general equations of a degree higher than four.

Evariste Galois was still only 20 years old when he was killed in a duel (1832). The night before his death at early light was spent outlining a comprehensive theory of the roots of equations (*Lettre à Auguste Chevalier*), the hastily scrawled origins of what today we call group theory.

From Galois' theory, mathematicians came to learn the importance of specifying the environment in which a statement is true or false—or perhaps entirely meaningless, and hence neither true nor false. It has been proved, by means of group theory, that a general equation of degree higher than four cannot be solved by radicals.

Radicals

The rational operations (addition, subtraction, multiplication and division) and the extraction of roots were the only algebraic operations known at the time when third- and fourth-degree equations were solved successfully. Thus, attempts to solve equations of higher degrees were limited to these elementary operations.

Galois showed that an equation is solvable by radicals if, and only if, its group for a field containing its coefficients is a solvable group.

Solvable groups

Solvability by radicals is possible if, and only if, the group itself is solvable. This means that, essentially, the group can be broken down into prime-order constituents. (Prime numbers are positive numbers greater than 1 divisible only by themselves and 1.) These constituents always have an easily understood structure. The term, prime-order constituents, is used because of this connection with solvability by radicals.

Perhaps, however, the heart of Galois theory—the theorem that an algebraic equation is solvable by radicals if and only if its Galois group is a solvable group—is merely a tributary result of Galois' true preoccupations. It would seem that Galois' contributions to creative culture were far more transcendental than the 'heart' of his mathematical theory.

A world of another paradigm?

Galois, was, in fact, much more a humanist and revolutionist than a professional mathematician of his time. Indeed, two outstanding mathematicians of the age (Cauchy and Fourier) rejected both Galois and his work.* Galois was expelled from his school, he was imprisoned for being a humanist, and he was 'framed' to fight the duel in which he was killed.

Galois needed mathematics in order to survive, but he was unsuccessful. The world according to Galois belongs to a different paradigm. One could even anticipate a number of indications that suggest a different level of complexity (as illustrated on page 71)—hidden, in a way—in Galois theory.

For example, finite fields with a characteristic of 2 (a Galois field of 2^n elements) have been used to design the best known codes for detecting and correcting automatically errors in the transmission of digital information.

It is possible to anticipate, through Galois theory, the tremendous effect that can be produced by a small cause, such as a slight change in a single postulate. This concept is indispensable in contemporary cybernetics.

One of the most complex and attractive problems related to Galois theory is the heuristic involved in the process of forming a transcendental Galois field and revealing (again, heuristically) the hidden structure of the field. This helps us to comprehend instantaneous communication, especially applicable in our understanding of the behaviour of subatomic

* And yet Baron Augustin Cauchy (1789-1857) went ahead to develop group theory, based on the pioneering work of Galois. Cauchy's efforts on the conditions pertaining to series convergence and his theory of the functions of complex variables helped bring a new rigour to the mathematical enterprise. Baron Joseph Fourier (1768-1830), another well known mathematical innovator, lived at first a stormy political life with, and then opposed to, Napoleon, 1798-1815. His elaboration of the Fourier series promoted him to a special rank among the 19th century's mathematicians.

particles. The very existence of instantaneous communication is very much under discussion by mathematicians and physicists in a number of countries, and the debate is a keen one.

Both categories of problems—related to field formation and the hidden structures of a transcendental field—cannot be considered at all by using today's mathematics. The two problems may be very much at the boundary of perception and communication that are far from orthodox and therefore are relevant to 'computer art.' Our research into Galois fields has enabled us (and us alone in the field of art, we believe) to investigate dematerialized existence. The exploration of dematerialized processes, existing outside space and time, is regarded as the most interesting aspect of research in computer art today. ■

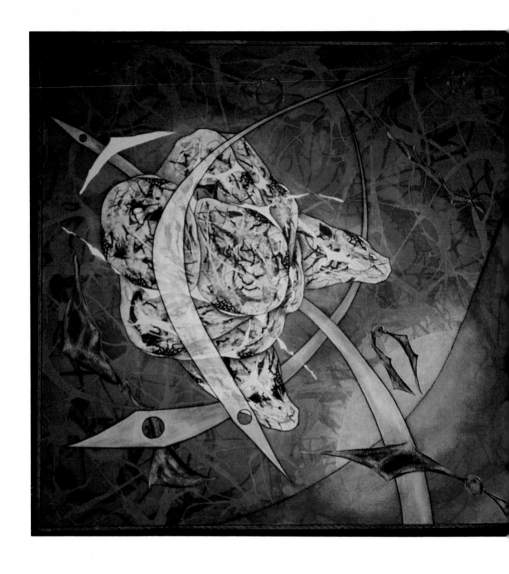

Illustration "An Affection" colour transparency by the author A. Kessler.

A painter trained in neurophysiology and surgery explains the motives and affective inspiration behind the works of art that he conceives and brings to reality.

Chapter 7

Innovative illustration
and the emotions

Alfred Kessler

Alfred Kessler is an American artist living and working in France. Born in 1922, he did his undergraduate studies at Fordham University, his medical studies at Duke University. He was trained in neurosurgery at the University of Chicago. As a young student, he painted but interrupted his artistic pursuits while completing his studies as a physician. He abandoned neurosurgery in 1962 in order to devote his full time to the practice of art. Dr. Kessler lives at 3 passage Rauch, 75011 Paris (France).

My paintings have nothing to do with theory, they do not involve reality *per se*, and they are not based on science. They involve human emotions in a fantasy situation that reflects the real. Fortunately—or unfortunately—we are all involved with man's emotional drive more than with any other human factor.

My scientific background (if one considers medicine a science) is obviously present in my paintings. The human form is the central theme of each painting. My background has much to do with the lesser aspect of the painting—the technique itself, which was developed solely to express better the image's essential meaning, its emotional content.

I aim for a transparency in each painting like that of a microscopic slide, but I do not depict anatomical relationships. Instead, I develop levels of abstract forms that are meant to give a force and meaning to what the human form is experiencing. The transparency has a twofold effect. Its clearness, as in a fine water colour, heightens the intensity of our colour awareness. Because of this, the transparency produces another and more important sense of seeing through or into the human form, approaching the depth of its emotional experience.

The emotions bind us together

My primary intention in painting is to represent a few human emotions that, in my belief, have not changed with time, style of life or the sophistication of society. My works deal largely with the emotions of love, loss, desire, fear: qualities that are common to all mankind and transcend time.

I make no attempt to portray specific features and try to avoid all signs denoting culture. In my illustrations you do not see a Frenchman, a Mexican, a Russian; you do not see a white person, a black one or an Asian. You do not see a good man or a thief. You do not know if my subject is man today, 50,000 years ago or 5,000 years from now. You see only man and woman in the ecstasy of their emotion, surrounded by a rather hostile environment—of which they are usually largely unaware.

My paintings are emotional, if not in their execution at least in their intention. Despite mankind's different ideologies, in spite of our minds and their diverging and frequently incorrect thoughts, our emotions are all the same. They unite us, if nothing else does.   ■

Note

The original illustration of *An Affection* was executed in acrylic paint on linen canvas. Colour transparency by the author.

Man's artistic enterprise has proceeded from simple, two-dimensional portrayals on cave walls of the things of nature to perspective throughout the plastic arts, from musical expression meant to record sights seen and sounds heard to that reflecting the inner self. This evolution has paralleled the increase in man's intelligence and his acquisition of knowledge, including know-how based in physics and mathematics.

Chapter 8

The growth of knowledge and the rise of art

Alfredo Planchart

Alfredo José Planchart Manrique is a doctor of medicine and a pharmacologist with long experience in physiological and nutritional research and in teaching. A prolific lecturer and author, Dr. Planchart is also a member of the Venezuelan National Academy of Medicine. One of his current interests is honesty and ethics in scientific research. The author was ambassador to Unesco, 1983-4, and has since resumed a limited medical practice. His address is Apartado 88742, Caracas 1080-A (Venezuela).

Man is a gregarious being. It is not possible to conceive of him as co-existent with solitude, not only because as the philosopher Ortega y Gasset wrote of the individual, 'I am, I and my circumstance', but also because biologically, man cannot be considered in isolation. Man is a social being; he needs other people, his fellow men. Society is basic to his existence. This dependence is due to his biological make-up; it might even be more accurate to say that it reflects his most important biological characteristic, which is the need to communicate, the need to belong to a community.

Although it remains debatable whether or not, before a child begins to speak, the nerve structure already exists for the creation of speech or the transformation of ideas into phonetic symbols—i.e. words—or into gestures, man obviously has a need to communicate. We should therefore determine and analyse what messages he communicates to his fellow men. Before making any epistemological classification, it should be noted that what man communicates is the result of the influence of the external world on his brain, conveyed not only through his senses but also through his own self-recognition. There is general agreement that man is the only animal that is aware of its own existence. Man's consciousness is also part of his surroundings.

It is not my intention to discuss the recent substantial advances in the physiology of thought made by great neurophysiologists such as Sperry, Wiesel, Changeux, Eccles and others, though the problem is far from settled. We should regard the brain as a 'black box' which receives stimuli and responds to them through movement; but we are not yet able to determine what happens inside it. What we are concerned with in this particular case is that it is precisely man's response to the external world which is transformed into a symbol. The symbol encapsulates the significance to the message of the idea formulated in the black box. The message is then emitted to the outside for the purpose of communicating with other human beings. This creates the unity, and consequently the culture, of the human species.

In a sense, the hackneyed and totally misunderstood idea of a superman is inconceivable as applied to the individual; it can relate only to a society, or to the entire human species. It is the intuitive perception that with the human species evolution is not physical but social. Viewed thus, superman would be the whole human race at the completion of its evolution; this would also be the goal of evolution. In this sense, the species as a whole is evolving to become a superspecies, and this explains man's need to communicate.

On genes, natural selection and morphology

The two factors needed in Darwinian evolution are the genetic component and natural selection or, in other words, the dual influences of heredity and the environment. These influences can be regarded as a struggle between, or a combination of, two contending factors, and the result is a compromise between them. The biosphere seems to be continuously seeking a solution to the problem of life in one direction or the other. Until the appearance of man, the influence of natural selection (that is to say, the environment) predominated. Animals could neither change nor dominate their environment. Hence the title of Darwin's book, *On the Origin of Species by Means of Natural Selection*. It was natural selection, the environment, that determined the course of life until the emergence of man.

The biosphere is thus dynamic. With the driving forces that constitute it, one might say that it seeks a Newtonian resultant, or 'solution'. The solution to date, which has perhaps been a successful one, is the human species. It represents the triumph of genetics, of life, over the environment, the emergence of intelligence as an evolutionary factor. In man the physical signs of evolution, though they have not completely disappeared, are not very evident. Any anatomical evolution since the appearance of *Homo sapiens sapiens* has been very slow and barely noticeable. The fact that physical evolution has come to a standstill has as its corollary that human evolution must necessarily be a cultural phenomenon.

Comparative anatomy shows that human morphology is foetus-like. It has not developed towards or in a given direction during its physical evolution. Other species, through their adaptation to the environment, develop only in a single specialized direction, whereas the human species does not have to specialize. Not being conditioned by their environment, humans must alter it to be able to exist. This sets man apart from other species, which are forced to be in equilibrium with their environment: for example, if the prey disappears, so does the predator. *Australopithecus* could come down from the trees, but he did not become master of the savannah until he had invented weapons to defend himself from his enemies and for hunting. Since he was omnivorous, he could adapt to a new type of food. His anatomy and physiology were not limited to development in a single direction. However, as he was not endowed with innate physical strength, he needed devices and instruments, which he developed with his intelligence and in co-operation with other human beings. For this he required a more sophisticated and detailed means of communication than the gestures which Darwin recorded as a means of communication among animals. He needed language, with a fuller and more highly developed symbolism, which became increasingly complex in order to keep pace with increasingly complex ideas.

It is true that communication exists at all levels of the animal kingdom, as established by N. Tinbergen and K. Lorenz, but the content of the message transmitted differs totally for the human race as compared with animals. For instance, animal messages cannot express abstractions. Communication in animals permits survival of the species but it is not basic to their evolution. That language is a product of evolution can be deduced from the fact that some characteristics of animal communication systems are to be found in it.

Transmitting the message

While it would be valuable to examine the characteristics of speech and language as a message, all that can be said here is that a study of speech and language, in the light of information theory based on the mathematics of thermodynamics, shows that they approximate to reality, in the sense of truth, in proportion to the volume of information they contain. In any message, information is dependent not only on the symbol used to transmit it but also on other factors such as the emitter and the receiver, and outside factors, such as 'noise', which determine the possibility of effective transmission. In other words, once an idea has been conceived in the brain, as a reaction to the external world or originating from the individual, the quality of the message depends on the person sending it, the precision of the language itself, the influence of the medium in which it is transmitted (physical or intellectual), the physiological or mental reception capacity of the person receiving it, and so on.

For the purposes of this study it is essential to classify the type (or types) of messages that contribute towards the evolution of the human species with a view to analyzing the influence of communication on cultural development. However, both culture and civilisation are circumscribed by geography, somewhat in the same way as an ecological enclave. Geographical location may well be the cause of the differentiation of cultures. Similarly, different environments may also act on different areas of the brain. Sperry's differentiation in general terms of the function of the right hemisphere of the brain as being to act on the abstract, art and culture, not only bears this out but also suggests the possibility that different parts of the hemisphere may act on different characteristics of the abstract, culture and art, which would constitute a physiological basis for the differences in cultural expression.

The fact that geographical separation has an effect on the message, which itself depends on the action of various parts of the brain lobes, might also explain why, as long as geographical separation existed, it was very difficult for cultures to understand each other. It is argued that with the development of communication and the narrowing of the gap, for example, between East and West, understanding improves. Such contact may lead individual members of different cultures to use new regions of their brains. It may be argued that when contact is established, 'noise', or irrelevant signals as defined in communication theory, is gradually reduced.

For our purposes, we might assume that the ideas produced by the external world are basic to communication, can be divided into two groups, the one quantitative, concerned with measure, and the other aesthetic, concerned with beauty and unity. The former are the source of science and the latter of art in Western culture. The source of the symbolism for both types of messages is the same. The foundation of Western knowledge is Pythagorean.

The heritage of Pythagoras spreads

Without wishing to embark on a discussion of the definition of art, we would simply note that we use the term 'art' here in its modern acceptation, in the sense of the fine arts, limited in particular to painting and sculpture, especially painting, because sculpture can represent, as does literature and music, abstract ideas. The Greeks were able, both in their mythology and consequently in their sculpture, to represent their vices and virtues typified by mythical characters. The Greeks could thus portray abstractions.

Freud later recapitulated these, applying the mythology so beautifully represented by the Greeks' statuary of the myths. Here, therefore, we confine ourselves to the visual art of painting—although Plato and Socrates included manual skills and even medicine in their concept of art, as also did Adam Smith.

The problem is thus reduced to the representation of the three-dimensional external world on a single, two-dimensional plane, which, when all is said and done, is what painting is. Basically, however, reducing three dimensions to two is a problem more akin to geometry than to art. Philosophy predominated in Greece; it would now be called epistemology, since it more closely resembled a philosophy of knowledge. The Pythagoreans were genuinely interested in investigating the physical world, and sought to solve theoretical problems. Leonardo da Vinci himself considered drawing fundamental for knowledge, and necessary for the analysis of visual objects. Thus from the outset the perception of our three-dimensional world, that is to say its representation in two dimensions, is no different from a purely scientific act, even if it seems solely artistic.

It should be kept in mind that for the Pythagoreans, mathematics meant everything that could be solved with ruler and compasses; in other words, the very concept of mathematics was governed by geometry, that is, the idea of form. Arithmetic came later, being more abstract. In view of the fact that recent developments in cosmological theory, such as relativity, are based on geometry, Riemannian geometry is an additional proof of the Pythagorean basis of Western culture, since the Pythagoreans were the first to find a logical, epistemological basis for the need to transmit ideas conceived in the brain. They used mathematical logic, and because they needed to transmit their concept of the external world, even the most comprehensive, *Gestalt* aesthetic feeling had if possible to be expressed geometrically. It was necessary to reduce the world to two dimensions, to encompass it in drawings.

It is obvious that this problem, and its developments throughout history, are Pythagorean in origin. The prehistoric paintings at Lascaux and in the Altamira caves do not reflect the need to copy the external world; in my view, they are not so much the transmission of a 'natural' message as an invocation. Their purpose is not communication with other human beings, but religious symbolism. It is not in these prehistoric paintings that we find the beginnings of pictorial art, or at least not the development of painting as a representation of the external world. Cave art may possibly be more closely related to the problems of modern painting, that of the twentieth century, which ushered in a new trend involving the search for a new goal in art, particularly in painting—one which is less concerned with the 'external world' as discussed above, and therefore more abstract, but with an affinity to prehistoric painting and the medieval primitives. This need for a new purpose in painting can be noted at other periods, for example with the Impressionists, Giotto and the primitives of the early Middle Ages or Dark Ages, or with individual artists such as El Greco, Goya and possibly Manet.

The celestial world is not three-dimensional

Perspective was thus a fundamental problem in the early history of painting. Greek culture was not able to solve or consciously analyse it, although from among the few relevant frescos of ancient times a mural has been preserved in the Uffizi of Florence, known as *Alexander at the Battle of Issus*, which suggests an attempt to solve the problem with the length and slant of the warriors' spears, indicating that the artists at least sensed the need for perspective. (It should be noted that in the fifteenth century, when the importance of perspective had already become known and artists were aware of the need for three-dimensional representation in two dimensions, Paolo Uccello, with the same technique and an almost identical subject, again attempted to solve the problem in the same way as the Greek artist; he used the effect of slanting spears, as can be seen in his paintings in the Louvre, Paris, the Uffizi and the National Gallery of London, on the subject of the battle of Florence.)

Thus, although relatively few fragments remain of Greek painting, what has been preserved shows that even when, from our point of view, the painting is an attempt to deal with an aesthetic problem or to express an idea through art, what is primarily involved is a system for the transmission of the effect that the external world produces in the mind of the artist (or artisan, as Plato would have argued); and the only solution was to use a methodology and a geometrical concept which were close to Pythagoreanism, and closer to science than to art.

Palaeo-Christian art is no longer the representation of the three-dimensional world, but of divinity, of the sky with all its heavenly hosts of angels and saints. Here we have the beginning of the period called the Dark Ages. Predictably, as the ideal is no longer Greek, art is no longer Pythagorean. The influence of Greek philosophy, of the epistemology that leads to the need to express oneself in quantitative measures and with beauty in mind, entirely disappears. The heavenly world is not three-dimensional. The problem of reducing three dimensions to two does not exist for the artist, and to transmit an idea conceived in his mind he does not use the same parameters as the Greeks.

It is possible that the success of Christianity in the social and political sphere was due to the need felt by the ancient world for a change in structures which would reform the situation of the social classes and bring about geopolitical and demographic development. Mankind could not continue to live in the structures of the Roman world. The reforms proposed by Caesar, which led to his assassination because of the rigidity of Roman politics, were for the same reason never carried out. But Greek culture's hold on people's minds did not permit the objectives of Christianity to be expressed in painting. The godhead, and heaven, cannot be reduced to the Pythagorean system of the ruler and compasses.

Thus for the primitives the problem was not a matter of perspective, and they therefore sought a new symbolism. The background of their paintings is always golden. Gold was the most precious substance, just as wine and bread were considered the most vital commodities of the time, and the golden background represented heaven. The nature of the personages was symbolized by the size of the figures: God was the largest, followed by the Virgin Mary, the angels and the saints coming after.

The innovation of 'comprehensibility' in art

Giotto's genius, in the early Renaissance, lay in the realization that the representation of abstract ideas or supernatural concepts could not be achieved by Greek-inspired pictorial art, corresponding, with its emphasis on the scientific, to the notion of art as a description of the external world, a concept in which there was still no distinction between aesthetics and scientific 'reality'. Giotto sensed that if he wished to continue in the Western tradition of Greek origin, which was one that his cultural background could understand, he would have to proceed in the same way as the Greeks, with his own mythology and his own artistic visual conventions.

In order to depict the Christian ideal, Giotto had to portray it in anthropomorphical terms, giving virtues and ideals a human morphology; he also had to make its symbols concrete, and paint God, his angels and his saints in a way in which they could be grasped by Western minds, as three-dimensional figures. He thus returned to the Greek objective of painting as a two-dimensional representation of a three-dimensional world. He painted God, his angels and saints as he imagined them. He could not conceive of heaven as anything but a three-dimensional space. Thus his series of paintings of St. Francis of Assisi, the depiction of the space between God and the saint in the *Stigmata* in the Louvre, and the blue cross standing aslant against the wall, all convey an impression of space. This is the beginning of the solution to the problem of perspective.

The solution itself came after Giotto, and was one of the great achievements of the Renaissance. Renaissance painting can be said to be 'comprehensible' thanks to the solution of the problem which had existed since the beginning of Western culture. That is why since the time of the Impressionists, and especially with modern painting, people say that they cannot 'understand' art; it does not comply with the normal standards of 'comprehensibility' which have prevailed in the West since the Greeks. The reason is that we are on the threshold of a new era, or another new era, in which a solution will have to be found to the problem of separating science from art as traditionally conceived hitherto.

With Giotto's anthropomorphism and the rediscovery of Aristotelian logic painting set out on a new course, closer to the Greeks than to Christianity, its objective ideals being the same. Painting again meant depicting man and his surroundings. It was obvious that there was no other way, and no other method of depicting them, than those traditionally regarded as 'comprehensible'. Once this principle was established, whether consciously or unconsciously, painting, although it still appeared to be the expression of aesthetic feeling, again became a technique based on scientific analysis.

Geometric harmony and perspective

In few Renaissance painters is this approach so obvious as in the great Piero della Francesca. His genuinely scientific concern for technique is clear in all his works, which deal with problems of perspective. It is said that he never undertook a work without studying it for days beforehand. Evidence of this can be found in *The Flagellation of Christ*, in which the architectural composition and the analysis of space are much more important than the flagellation itself; the essence of the work is not the religious motive but the study of perspective. In his *Madonna with Pearl*, it is clear that while, as was normal for the period, the subject-matter was the Christian ideal, the artist's concern was the representation of the three-dimensional world. This is evidenced by the fact that there is no account of the Madonna with a pearl in religious tradition, nor any other similar portrayal of her. The pearl is egg-shaped, and the Madonna stands in an elliptical recess, so that the artist's problem is that of painting an ovoid shape against an ellipse, i.e. a problem of perspective as opposed to mere projection, a mathematical notion involving abstraction, hence a revival of the Graeco-Pythagorean approach, which is unmistakably scientific.

Pure science, as we now know it, began to separate itself from painting. It is relevant to note that the discoveries of Galileo were not far off, and that physical science, slumbering since Archimedes' time, began to experience a rebirth.

That great figure of the Italian Renaissance, Raphael, showed his direct interest in the problems of perspective in paintings such as *The Marriage of the Virgin*, so similar to the work by Perugino in the Sistine Chapel. The mosaic floor is simply an exercise in seeking the vanishing point, that is, a problem of pure perspective. Albrecht Dürer made direct studies of perspective, and his engravings which project both landscapes and human figures on a single plane are known the world over.

It is not surprising that Leonardo da Vinci, Brunelleschi and a number of his contemporaries considered Alberti not only the greatest artist of his time, but also a great painter, although none of his pictorial works are known; the reason is that Alberti, like all good architects, carried out theoretical studies on perspective in painting, which were of fundamental value to Renaissance artists.

However, the separation of art from science did not really begin until perspective became established as a distinct branch of geometry, the projection of figures on a single plane, leading up to descriptive geometry.

With the outstanding figure of René Descartes, mathematical analysis reached the highest point of the Greek ideal of the abstract expression of the external world, marking the extreme difference between human and animal intelligence, namely man's capacity to conceive abstract ideas. Descartes succeeded in interpreting form through symbols, substituting algebra for geometry. Thanks to his mathematical analysis, a series of letters and signs could now express everything that a three- or two-dimensional figure could express.

A new flowering of intelligence

This was the time when science and art went their separate ways. Their two systems could now be approached differently. Bodies in space, which could henceforth be dealt with by physics, no longer needed to be two-dimensional figures, which could be expressed algebraically. Problems of movement and force could now be solved algebraically, a mathematical symbol sufficing to express a concept of physics. This line of thought was continued by Newton, and culminated with Leibniz and his differential calculus, the maximal degree of abstraction, the apotheosis of Greek thought. Newtonian mechanics, as subsequently evidenced by the development of the theory of relativity and the quantum theory in physics, sets aside reality and works only with ideas, admittedly with a greater amount of information than in previous systems, but less than in modern physics. Modern physics distances itself from the Greek idea of perspective and Euclidean geometry; but in its quantity of information it more closely resembles what could be called reality, although this reality, or 'truth', is not represented by it or recognized as operative, since the abstract or Euclidean ideas of truth and reality are not considered applicable in modern physics—a possible further departure from the Pythagorean system. As human intelligence develops, it constantly seeks new courses.

The flowering of the Renaissance, the development of the message, the triumph of the new system, also led to a turning-point in painting. Even though El Greco can be said to be a follower of Tintoretto, traces of whose influence can be found in his work, one can see that his approach is not the same, as if he did not agree with the Pythagorean system and did not feel the need to copy the three-dimensional world. Pure aesthetic feeling can be sensed in his painting, drawing him closer to Altamira or Lascaux; he painted with feeling, with evocation. As said until recently of J.S. Bach, it is art for art's sake. It could also be said that El Greco achieved the mediaeval mystic ideal more profoundly than pre-Renaissance artists. Here we have perhaps the beginning of the separation of art from science.

It is possible that Mannerism, in its pursuit of the sculptural ideals of Michelangelo, also failed (with the exception, perhaps, of Guido Reni) for the same reasons as classical painting, continuing on Pythagorean lines, in its attempt to solve an already solved problem, that of form, the expression of the three-dimensional world in two dimensions. The separation of mathematics, and consequently physics, from the problems of pictorial representation meant that the three-dimensional world now became the subject of investigation of these sciences, together with movement and force, with which painters and artists were not concerned.

It is now clear that the great post-Renaissance painters, such as Rembrandt, although their goals were basically aesthetic expression, did not show a complete separation between art and science. Both Rembrandt and Caravaggio, with their chiaroscuro effects, showed an awareness of the physical problem of light, a physical component of the external reality, which was only to be solved three centuries later by the Impressionists.

Form and decoration recede in importance

It may well be that this new turning-point in painting, which might also be called Pythagorean, and which reached its conclusion in the nineteenth century, was again linked to the fact that the problem of form was no longer solely pictorial, but that there was another dimension to be expressed. In Watteau the need to escape the indoor world of drawing-rooms can already be felt. Until that time landscapes had been merely decorative. In the *Mona Lisa* and *The Virgin of the Rocks*, the landscape accompanies the figure as a decoration, but it is a real landscape, no longer in gold as in primitive paintings. It remains however a decorative background, never a central motif, it simply provides additional information. With Watteau, landscape plays a greater part: even when the personages seem to dominate, their setting is of vital importance.

This trend continued with the heroic figures of David, Ingres and Delacroix, culminating with Corot and Courbet; all of them were saying that a way must be found of expressing light, and that it is not only form which conveys visual information. It was, however, Manet who found the key to the problem, although he could not have done so without the genius of Goya, who, like Giotto and El Greco, stands as a landmark in the development of art. These artists created art for art's sake. Goya's inner world resembles that of Beethoven, the expression of his own self rather than the solution to a problem. In the author's view Manet could not have experimented as he did with light if Goya had not existed.

The clearest illumination of the dead end to which Pythagorean painting had come can be found in the Venetians, Canaletto and Guardi. While it was with Titian that the summit of Renaissance art was reached, with Canaletto and Guardi, the system based on form had already expressed everything it could. Perhaps a similar phenomenon occurred in music with opera. After Wagner, Verdi and Puccini, another approach had to be found, since the one they adopted had come to a dead end. This might explain why operas of the same quality have not been written since.

Impression and the problem of light

Manet realized that light, in the sense of colour, is capable of expressing what cannot be obtained through formal perspective. There had been previous attempts to create the feeling of volume using this method. El Greco, for instance, with his colour contrasts, created a greater sense of space.

Form and colour are complementary in creating visual impact. Our appreciation of black and white, of the information provided by the various shades of grey, is artificial: it comes from prints, and especially photography. I know of only one picture from the classical period in 'black and white' apart from etchings; it is an oil painting by an unknown painter. In Venezuela there is also to be seen an experimental oil portrait in black and white, by Manuel Cabré.

While the problem of form, the reduction of a volume to a single plane, apparently solved the problem of representation of the external world, so that the communication of information seemed to have been achieved, and only aesthetic feelings remained to be expressed, something else was felt to be missing, something that could not be expressed in geometry. It was still a scientific problem, for physics is not merely kinetics and form; there remained the problem of light. The Impressionists were the ones to find the solution. The advent of Impressionism in France is now considered by many to be the greatest achievement of expression in painting. However, the Impressionist approach should still be regarded as Pythagorean. It represents a return to the physical expression of the external world, as if returning to its Greek origin. There is great aesthetic expression, but the information is transmitted through a scientific process—light.

In England, Turner had already depicted the importance of light and movement in his admiration of the forces of nature. It is not possible to set aside the influence of this great artist before we proceed to the Impressionists.

Manet should be regarded as an intuitive forerunner of Impressionism rather than as a true Impressionist. Admittedly by his technique of foregoing draughtmanship in preparing a painting, and using only colour, he introduced light; but in this respect he was closer to Van Dyck, who in his time had already discovered how to reproduce instantaneousness of expression. One might call him the Van Dyck of Impressionism. His rapid solution to the problem of background and the portrait, putting greater emphasis on inspiration than on detail, singles him out as an authentic precursor.

Communication contributes to evolutionary progress

It was not until Monet concentrated on the problem of light in his *Effects of Light on the Thames*, in the Marmottan Museum in Paris, and particularly in his *The Houses of Parliament in London*, that the importance of form replaced the physical effect of light on the retina: here was a new Pythagorean vision of the external world. It is with these investigations by Monet, not as previously claimed Manet's *Déjeuner sur l'herbe* or his *Olympia*, that Impressionism truly began. Perhaps the greatest unconsciously Pythagorean interpretation is *Le Moulin de la Galette* by Renoir, in the Jeu de Paume Museum, Paris. There is not only the effect of the contrasting colours of the figures and trees which give the impression of the outside world and perspective, but also the rays of the sun filtering through the leaves which can be felt and sensed in the air, and are reflected in the objects and faces of the personages. It is a solution to a physical problem of perception: the physics of light in the form of images.

The other Impressionists followed this approach and reached their own solutions. Degas did the same thing with his dancers; Pissarro contributed the light of the tropical sky; Sisley, Berthe Morisot each had his/her word to say. The search for a solution explains the flowing of so many geniuses at once. Yet at the height of the movement, it was sensed that new paths and new approaches had to be sought: for example Seurat, with his pointillism, the great Cézanne, with his use of colour and relatively limited use of perspective, and others; until Van Gogh appeared with the beginnings of Fauvism and his succession of paintings which reveal the development of his mental state.

Fauvism, continued by Vlaminck and others, is the beginning of another turning-point in history. Again, Impressionism found another way to communicate, representing the external world in perspective, but in a perspective which included light.

The work of Picasso, Matisse, Derain and the painters who followed them shows a deep concern for a new solution to the problem of expression. Now that everything which could be said about and with light had been expressed by Impressionism, a new solution was required to the need to communicate. Human beings had received a view of the external world, but an inadequate one, conveyed only by the retina and the optic nerve. As evidenced in anatomy and physiology, the function of vision cannot be obtained with a single organ, the eye, or with a single portion of the brain. While the right hemisphere of the brain may be said to influence the concepts of art and abstract ideas, this is far from being the whole problem. Also involved is the integrative function discussed by Sherrington, and in fact the integrative function not only of the entire nervous system, but also of society. What we are concerned with is the progress of evolution through communication among individuals. The problem is one of the evolution of mankind as a whole.

To delve more deeply

CHANGEUX, J.P. *L'homme neuronal*. Paris, Fayard, 1983.

ECCLES, J.; POPPER, K. *The Self and Its Brain*. Berlin, Springer, 1983.

KLINE, M. *Mathematics, A Cultural Approach*. Reading, Mass.,
    Addison-Wesley, 1962.

PLANCHART, A. El concepto de la verdad en la ciencia actual, *Acta Cientifica
    Venezolana* (Caracas), 1965.

SPERRY, R. *Science and Moral Priority*. New York, Columbia University Press,
    1983.

TSUNODA, T. *The Japanese Brain* (trans. Y. Oiwa), Tokyo, Taishukan
    Publishing Company, 1985. The author, an audiologist and otologist, contends
    that certain neural perceptions among Japanese are seated in other parts of the
    same brain hemisphere or else in the opposite hemisphere than is the case
    among non-Japanese.

A specialist in the history of music traces the mathematical influences on the development of, principally, Western music over the last millennium. These mathematical influences sometimes have their origins in nature, and the relationship between natural phenomena and musical expression have influenced in their turn the affective nature of music. The author also explains the genesis of various scales used in musical composition since the ninth century of the Christian era, emphasizing the close relationship between science and art when innovative forms of expression are sought.

Chapter 9

Geometry, arithmetic and musical creation

Nils L. Wallin

Nils Lennart Wallin studied musicology, philosophy and ethno-anthropology at the universities of Uppsala, Stockholm and Basle, and musical theory, violin and chamber music at the Schola Cantorum in Basle. Dr. Wallin has taught music and the history of music since 1951 and also served as music critic of the newspaper Aftontidningen *for six years; he was general director of the Swedish Institute for National Concerts (1963-70), then general director of the Stockholm Philharmonic Orchestra and the Stockholm Concert Hall (1970-76), scientific secretary of the Royal Swedish Academy of Music (1978-1982), and executive secretary of the International Council of Music (1982-1985). The author has also spent the past ten years studying the cognitive implications of new brain research on music. Address: Royal Swedish Academy of Music, Blasieholmstorg 8, 111 48 Stockholm (Sweden).*

Does the sun ring and peal?

We read in the prologue to Goethe's *Faust*,

> Die Sonne tönt nach alter Weise
> in Brudersphären Wettgesang,
> und ihre vorgeschriebne Reise
> vollendet sie mit Donnergang.

or 'The sun rings and peals in an honourable and well co-ordinated rivalry with the planets, completing its pre-ordained course with the crash of thunder'.

Goethe's vision is magnificent, his creation filled with force and joy. Still, one senses that this poetic image represents a very late and only metaphoric offshoot of an ancient intellectual tradition in cosmology. Goethe was surely keeping pace with the scientific knowledge of his time, but during the nineteenth century the cosmos did not ring and sing. On the contrary, outer space represented the ultimate silence.

Only about 200 years earlier, one of the pioneers of modern empirical science, Johannes Kepler, had filled his most important and famous work with creative statements concerning the idea of music as the co-ordinating force of the cosmos. His five volumes on world harmonies (*Harmonices Mundi*, 1619) described the behaviour and correlations of the six planets then known. Kepler's cosmic view is a theory on the harmonically co-ordinated movements of the planets. Kepler was the first to begin with the assumption that the orbits of the planets about the sun are elliptical rather than circular. The extreme points of the ellipse are at the greatest (*aphelion*) and least (*perihelion*) distances from the sun.

Even when planetary orbits change, their angle to the sun remains almost unchanged. This angle, for Jupiter, for example, is an aphelion of 4'32" and a perihelion of 5'30" (estimated by Kepler at 4'30" and 5'30" respectively); the angle corresponds roughly to a relationship of 5:6, or a 'small third'.

In much the same way, each planet was allotted a value describing its relation to the sun and—simultaneously—corresponding to the mathematical expression of the elementary musical consonants. Kepler even stated, 'Pour air in the heaven, and a real and true music will sound. There is a *Concentus intellectualis*, a "spiritual harmony" that gives pleasure and amusement to pure spiritual Beings and in a certain sense even to God himself, not less than to the man with his ear devoted to musical chords'. Kepler's interpretation of his astronomic discoveries, still partly valid today, results in a heliocentric-musical philosophy of nature; this is rather distant, however, from its historical origins in the music theory of ancient Greece.

Scotus Erigena and early Western polyphony

Between the ancient Pythagorean doctrine on harmonious co-ordination of the movements of planets (both among themselves and in relation to the earth), all manifested in the numerical order of the consonants, and the musical fascination of Kepler and G.W. Leibniz[1] combined with modern scientific thinking, there is an unbroken tradition of almost two millennia of cosmological thinking in musical terms. One recalls the names of Ptolemy, Boethus, Thomas Aquinas and St. Augustine. To illustrate just a small fraction of this very long bridge between classical thinking in music and the modern world, Johannus Scotus Erigena (who lived c. 800-880, during the Carolingian renaissance), may serve as an example.

For Scotus Erigena ('Scotus' refers to his origins in either Scotland or Ireland), music was part of the seven free arts, whose quadrivium included arithmetic, geometry, music and astronomy, and whose trivium embraced grammar, rhetoric and dialectics. Arithmetic held a central position in the series that Johannus certainly regarded as the natural functions of the mind—the foundation of the whole quadrivium. Music he defined as 'that discipline which in the light of reason contemplates the harmony of all things, whether they be in movement or in a status of discernible perseverance, and combined in natural proportions'.

Johannus' cosmological view is not heliocentric like Kepler's, but mainly geocentric. Farthest out is the firmament and its fixed stars, moving round the earth in twenty-four hours; the earth itself is motionless in the middle of the universe; the sun, between firmament and earth, moves in a circle within lower space; finally, Jupiter, Mars, Venus and Mercury circle about the sun (Fig. 1). From earth to sun, there is one octave, and from sun to firmament one more. The distance between earth and moon corresponds to a whole-tone interval, that is 18:16 or nine-eighths. So in Johannus' universe one tone and one sphere correspond to each planet and the firmament beyond: 'the ethereal circle turned the starry sky, which went in circles round the world. In manyfold advance the consonant crowd of planets moved, emitting sweet whole-tones, six in number, [with] seven intervals and eight tones'. This makes a complete diatonic scale.

The harmony of heaven as well as of the world thus depends on mathematical proportions. The cosmic harmony has the same design as the musical one, in the sense that it is composed of different parts arranged with respect to mutual relations and proportions.

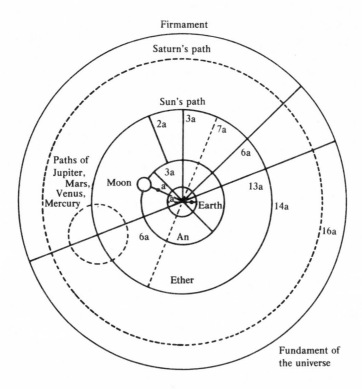

Fig. 1. Scotus Erigena's geocentric view of the cosmos. In this universe one musical tone and one sphere correspond to the planets and the firmament beyond. (From J. Handschin, Die Musikanschauung des Johannes Scotus (Erigena), *1927.)*

I realize [wrote Scotus Erigena] that the unity of the whole nature depends on different and, in respect of each other, even opposed things. I have become aware of this by musical proportions; thanks to them I learned that only the rational relationship between different musical intervals which, connected with each other, bring about the sweetness of music, can please the mind and create beauty .... Even if the listener might not understand what is the cause of the sweetness and the beauty of the tones, still everyone has an inner sense for proper combinations and relationships of different things.

Intervals, harmony and polyphony

Here Scotus Erigena touches the notions of *Gestalt* theory but, also, the seemingly eternal problem of how to explain the perceptual quality of consonant in a musical context. This remains not satisfactorily explained today, despite the hypothetical critical bands of psycho-acousticians and the receptive fields devised by neuroscientists. Jacques Handschin has pointed out that Johannus' 'inner sense' may come close to the suggestion made by Leibniz's pupil, the mathematician Leonhard Euler, that the grade of consonants and intervals has a direct relation to the grade of simplicity of the multiples of the two numbers expressing the interval, such as 1:2, 2:3, and so on. Euler arrived at some results which do not correspond to our experience; on the other hand, some recent neuroscience and other biological ideas show, in a promising way, data which may be interpreted as a kind of 'inner counting', where 'experience' (as used in cognitive psychology) is insufficient.

Scotus Erigena's way of discussing intervals indicates that he was acquainted with the new and progressive music of this period of the Middle Ages—the polyphonic setting of melodies. Polyphonic structure,[2] the distinctive feature which more than any other would set apart Western music when contrasted with that of other cultures, needs rational systemization of its tonal material. This is possible only through the use of sophisticated mathematical and physical knowledge, knowledge made possible at that time by the quadrivium.

Scotus Erigena probably knew the most important polyphonic document of the ninth century, *Musica Enchiriadis*, and its commentary, *Scholia Enchiriadis*. The background of the main tract is not clear. Much of its content is primarily an example of mediaeval speculative thinking, but the musical examples illustrating the doctrines maintained in the document are very good indeed; they are typical examples of ninth-century polyphony. Here speculative theory is no more distant from musical application than Rameau's *Traité de l'harmonie* is from his own compositions 900 years later or Schoenberg's twelve-tone rules from his Variations for Orchestra, Op. 31, published 1,100 years after Johannus.

The passage in the *Scholia* in which the position of music is established among the sciences indicates clearly the proximity of music to Johannus' conception of the world:

Pupil: How was harmony born of arithmetic as from a mother?
             And what is harmony, and what is music?
Teacher: We regard harmony to be a mixed symphonia of different sounds.
             Music is the theory of the symphonia itself. And because it is
             altogether dependent on the theory of numbers, like all the other
             mathematical disciplines, it is only through numbers that we
             understand it.
Pupil: Which are the mathematical disciplines?
Teacher: Arithmetic, geometry, music and astronomy.

Pupil: What is mathematics?
Teacher: A doctrinaire science.
Pupil: Why doctrinaire?
Teacher: Because it deals with abstract quantities.
Pupil: What are abstract quantities?
Teacher: Abstract quantities are those embraced only by the intellect because
             they lack material, i.e. physical admixture. And further: multitudes,
             magnitudes, their opposites, forms, similarities, relations and many
             other things . . . change when connected with physical substance.
             These quantities are each directly treated in arithmetic, in music, in
             geometry and in astronomy. It is thus because these four disciplines
             are not skills of human invention but important researches in holy
             works; and they support in the most wondrous way acute minds in the
             understanding of Creation.

Some rhythmical relationships

The true understanding of Creation was to Johannus, as it was to the
unknown author of *Scholia Enchiriadis*, knowledge of the inner qualities
of numbers. Belief in the *divini numeri* made mediaeval science a
mysterious mixture of arithmetic and musical knowledge in an ancient
rational tradition combined with Christian religious speculation. The
mixture was still at work in J.S. Bach's religious compositions, even in
creations by Hindemith, Schoenberg, Stravinsky and perhaps those by
Stockhausen, Peter Schat, Ingvar Lidholm and Xenakis—all so different
(see Fig. 2). In the writing of Scotus Erigena, the numbers 6, 7 and 8
represent perfection; they serve as multipliers in his astronomical
measurements and they represent the most perfect consonant, the octave,
the seven intervals and the six whole tones. Eight is the number
representing the age of the world, as well as a symbol for the unity of the
overlapping characteristics of the human organism (the body itself, life's
impulses, sensuality, reason, intellect). These continue at three higher
levels: the transition of the mind into knowledge; the passage of
knowledge to wisdom; and, finally, the fall of the purified mind to God.
In this way, the three-dimensional, perfect *soliditas* of the number 8 is
reached.

  That perfection was symbolized during the early Middle Ages also by
the number 3 is well known. As suggested above, the new polyphonic
setting of melodies necessitated new methods for the structuration of
tonal material. In *Musica Enchiriadis*, the rhythmical relationship
between Gregorian melody and its new counterpart was probably still
quite simple; the free, declamatory flow of Gregorian melody was
supported *punctus contra punctus* (tone against tone) by the added,

Fig. 2. The zodiac of the twelve tonalities, a mysterious mixture of arithmetic and musical knowledge, continued in use until the time of Bach and even composers of the early twentieth century. This 'tone clock' was developed by the Dutch composer, Peter Schat, to demonstrate the relationships among the different pitches.

second, part. This was possible so long as the tonal relation between parts remained simple, i.e. moved in parallel octaves, fifths or fourths. But when composers began to increase the structuration of the psalm (or song), by liberating for instance the melodic web of parts from strict parallelism—what eventually meant using simultaneities other than the octave, fifths and fourths, and by reciting the text differently in each part (or even by using simultaneously two different texts)—it was at this moment that rhythmical flow had to be regulated.

The emergence of new musical structures

During the late twelfth century a strict metrical, or modal, system was established, based on the triple metre in various combinations or modes. It took a very long time before this system was replaced by a rhythmical one, where the free flow of triple and duple metre was recognized within an enormously subtle and sophisticated mensural system. This became the basic building-stone in the growth of later medieval and Renaissance music. Every step of the kind in the ongoing structuration of music meant new conceptualization, almost always by means of arithmetic and geometry. Yet these were not always declared as desirable aesthetic innovations. The direct connection with 'holy works' remained unbroken; and the duple metre had to be avoided because it had a beginning and a middle but no end, and was thus imperfect.

This queer mixture of a strongly progressive and intellectually rational structuration, on the one hand, and, on the other, a need to motivate listeners and explain the steps taken in composition in metaphysical terms has often characterized musical thinking throughout history; our own century is a good example of this. The antinomy is illustrated interestingly in the person of Johannes de Muris, a mathematician at the Sorbonne who flourished early in the fourteenth century and a spokesman of what is still called *ars nova*. Muris, the Pierre Boulez of late mediaeval times, drew up in effect the programme of contemporaneous music in the treatise titled *Ars Novae Musicae* (1319), in which he developed ideas regarding music and its temporal progression; these thoughts come very near to those expressed by Stravinsky in 1946—or those that the scientist Victor F. Lenzen stated in 1938 in the *International Encyclopedia of Unified Science.*

Departing from the concept of trinity, Muris concluded that the number 3 is perfect, while 2 is imperfect and thus inferior:

But each compounded number derived from these numbers can duly be regarded as perfect, because of its similarity and conformity with the number of three. For time is—because it belongs to the class of continuing, continuous things—not only divisible with the number of three but also endlessly divisible with infinity [*sic*] .... Sound is created by movement because it belongs to the class of things which follow directly on each other .... Two successive phenomena do not exist without movement. Time indivisibly joins movement together. The necessary conclusion ... is that time is the measure of sound.

In 1946, Stravinsky stated that

the musical phenomenon is nothing else than a phenomenon created by mental activity. There is nothing in this suggestion which should frighten you. It just implies that behind . . . musical creativity there is a preceding search or a desire at first to move in the abstract, with the intention to give form to a concrete matter. The elements with which this . . . is concerned are *tone* and *time*. You cannot imagine music without these two elements.

The Pythagorean theory

We all know the story—told by Nichomachus almost 700 years after Pythagoras' death—of how the Greek mathematician discovered the relation between a given vibration and the pitch of its sound. Pythagoras had noticed, when passing a blacksmith working with hammers of different sizes, that the sound from the anvil changed depending on the hammer used. When he checked the weights of the hammers, Pythagoras found that they were related to each other proportionately, such as 1:2, 2:3, 3:4, and so on, and that the resulting pitches were the octave, the fifth, the fourth, etc. According to Nichomachus' narrative, Pythagoras continued his investigation by hanging on strings of equal length loads corresponding to the hammers; he stretched the strings in function of the weights, and he then transformed the results into instruments such as the aulos, a double-reed oboe, and the monochord, a long, wooden resonator having one string and a movable fret.

Nichomachus' tale is a legend; it does not correspond to either historical or scientific exactitude. What Pythagoras discovered when he compared pitches with different lengths of the monochord's string was that pitch—more correctly, frequency—is inversely proportional to vibrating length. We cannot be sure, however, that the quantity which Pythagoras and his fellow philosophers associated with tone was indeed frequency. There is much to suggest that the basic quantity of tone appeared to these early scientists as the rapidity of the tone's onset; we know that the first impulse of a tone tells us its pitch, that is, the 'height' of the tone.

Euclid spoke of 'more frequent movements' of the higher, and of 'more seldom movements' of the lower, tones; he meant that a tone consists of parts. It is a remarkable fact that the early Greek philosophers, being simultaneously musicians and mathematicians, portended modern science in this respect. In 1952, Archie R. Tunturi showed in experiments now classic that it is the onset of tone—its first 1-2 milliseconds, and thus not the entire wave—which makes possible our discriminating sensation. Complete perception requires more time.

The Pythagorean way of thinking was obviously more mathematical than physical in character: number was considered to be the essence. Pythagoras arranged the intervals he obtained from the division of a vibrating string in a diatonic scale framed by the octave 1:1-1:2. He arrived at this scale, or *harmonia*, by piling fifths upon each other, afterwards transposing them into a one-octave scale. The basic consonants—the prime, octave, fifth, fourth and their equivalents—are represented by integrals and their multiples, the exponents of the real Being. Harmonies, therefore, were also considered to mirror the essence of Being, including heaven and the relationship of its planets. Everything in Pythagorean philosophy circled about number and tone as its sensual apparition. Each number had a structure, manifested in and by a tone belonging to a scale. In this respect, Jacques Handschin has reminded us that 'among the different kinds of co-ordination that we can think of, the co-ordination with numbers is the most exquisite one'.

Even if Greek musical theory was an integrated part of ancient arithmetic and geometry, without a primary bearing on musical composition and performance, it has had immense importance in terms of creativity in Western music: as a way of rational thinking about musical functions and as a building block in musical ideology.

On different temperaments

Pythagoras' scale is not identical with the scales in general use in Western artistic music since about 150 years ago. We have seen that Pythagoras arrived at his scale by piling pure fifths one upon the other; this he did by dividing the monochord's string into two lengths relating to one another as 3:2, thus forming a 'circle of fifths'. If the notion of circle is correct, we should be led back after twelve steps to the initial tone but in another octave. This does not occur, however, so the image of a circle should be changed to that of a spiral since the twelfth fifth is higher than the first tone. This amounts to a difference of about one-eighth of a tone, yet this 'Pythagorean comma', small as it is, is mighty enough to make some musical operations unacceptable to the ear. Thus some melodic and harmonic progressions have to be avoided, as well as some combinations of keys.

Such tonal disadvantages were not revealed, of course, until a much later stage in the development of music. It was Western polyphony that, at a certain level of its structuration, made it necessary for composers, as well as designers of keyboard instruments, to find a solution to the problem. Compromise methods were tried, wherein the 'inaccuracy' was not concentrated in a few tonal and key variants but spread equally over all imaginable variants. These systems of compromise are known as temperaments, in which the intervals deviate from the acoustically correct proportions and values such as are found in the Pythagorean scale.

The most consistent solution in the Western tradition is the equal temperament; this is usually ascribed to Andreas Werkmeister, at the beginning of the eighteenth century, but its groundwork was laid during the sixteenth century by H. Grammateus and V. Galilei. The latter, father of the great astronomer, recommended the use of a half-tone having a frequency of 18/17ths—very close to the equally tempered one, and instead of the Pythagorean 256/243rds.

To make all this more tangible, let us use the expression of cents, a logarithmic measurement devised about a century ago by the mathematician, A.J. Ellis. The cent is a scientific tool meant to facilitate the comparison between intervals belonging to different scales and temperaments. Ellis used the semitone (the tone at half-step) of the 'equally' or 'well tempered' scale as standard; the cent is thus one-hundredth of the semitone. With the semitone of 100 cents, the pure fifth equals 700 cents, the pure octave 1,200 cents. It is easy to illustrate, by means of this system, the relationship between intervals belonging to the most different musical systems independently of age and culture. Figure 3 illustrates in cents the difference between the chromatic scales according to the tempered and Pythagorean systems.

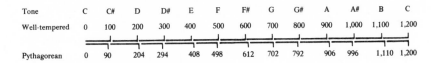

Tone	C	C#	D	D#	E	F	F#	G	G#	A	A#	B	C
Well-tempered	0	100	200	300	400	500	600	700	800	900	1,000	1,100	1,200
Pythagorean	0	90	204	294	408	498	612	702	792	906	996	1,110	1,200

Fig. 3. Relationship between notes according to the well tempered and Pythagorean system.

It is of interest to compare Figure 3 with the intervals belonging to the scale of another highly sophisticated music, that of the Javanese *slendro* scale. In this Indonesian scale, the octave (1,200 cents) frames five intervals of equal size, 240 cents: at the 240, 480, 720, 960 and 1,200 marks.

Metamusical thinking

What have the speculative ideas and research described above provided for the future of creative and innovating force? First, perhaps, we should not make too systematized or exhaustive simplifications of how the pondering of numbers in a tonal context inspired creativity and metamusical imagination throughout the ages. Behind even the intervallic speculations of the Greek philosophers, extremely old ideas from the megalithic period on the creation of the world can be traced to a first and spontaneous cosmic sound. The open jaws of the beasts adorning old church roofs remind us, too, of this notion of sound at the Beginning. In the cycle of life—the movement of time in minutes, hours, days, nights; the cycles of the moon and sun; the ages of man; the resurrection of life from death; wakefulness and sleep—the changes were sometimes symbolized by various animals. The bull was fertility, but also morning and spring; the ox was the gelding, and weakness, evening, autumn, surrender. The animals, like the conceptions which they illustrated, or really were, were then transformed into tones and intervals. The intervals were each given a specific ethos, via a non-musical thinking introduced to a musical system. Thirty years ago, Marius Schneider demonstrated in his work, *Singende Steine*, how the capitals of pillars adorned with heads of animals could be translated into a series of tones to form well-known hymns. The capitals arranged around the yard of a cloister, for example, served as singing pillars, what Dieter Rudloff has called a 'hidden musical notation'.

In similar inspiration, the architects of some of the great mediaeval cathedrals designed the length, width and height of the holy site according to the proportions of numbers characterizing the main consonants and at the same time the first harmonics. In this sense the cathedral of Chartres, built on a megalithic site, is truly frozen music. It is fascinating to consider the quality of intellectual or spiritual integration represented in such a cathedral, with its 'harmonious' construction co-ordinated with the 'harmonious' oscillations of music sounding from within the edifice.

Great artists of more recent times have built their works on this intellectual tradition. Hindemith, in his opera *Die Harmonie der Welt*, portrayed Johannes Kepler pondering deeply over the mysterious relations between *musica mundana* (the inaudible music of the cosmos), *musica humana* (music as a divine theory) and *musica instrumentalis* (the audible music of man, the actor). Hermann Hesse, in his novel *Das Glasperlenspiel*, created an élitist spiritual republic contrasting strongly with the present, commonplace or vulgar world. The main activity pursued in this utopia was a game, the rules for which were taken from musicology and mathematics; the rules could 'express the contents and results of almost all sciences and put them in relation to each other. The Glasspearl Game is, further, a game having all the content and values of our culture'. Hesse made a critical evaluation of the dualism between the purity, clarity and fragile abstraction of musical and mathematical conceptions, on the one hand, and the corruption, compromise and vitality of the qualities of everyday life, on the other hand.

From the late Middle Ages to dodecaphony

The mediaeval aspiration to found musical structure on melodic and rhythmic values expressible in numbers—representing harmony in a mathematical and spiritual sense—was, in a way, an attempt to integrate the musical structure with a single parameter, to illustrate the ultimate identity. Heaven and hell, life and death—the structure sometimes reaches such a complexity (in, for example, the isorhythmic motets of the fourteenth century) that the fine structural details are not audible in conscious perception. But what about the unconscious effects, the secrets of music?

We have already noted Leibniz's definition of music as the mind's secret art of counting. And we are beginning to grasp that there are within the organism several positive and negative feedbacks at different levels of biological complexity; these, co-ordinated, may be the background to this musical 'counting'. According to Kepler—and it is an astonishingly progressive and modern idea—the auditory system developed on the basis of mathematical relations pertaining to the physical world. Ideas such as these were elaborated by Thomas Mann in *Dr. Faustus*. His hero, the composer Adrian Leverkühn, makes a pact with the Devil, not how to obtain the secret of making gold, but how to resolve the enigma of identity. In a large musical composition, which Mann describes in detail, Leverkühn finally uses the same tonal material in describing both heaven and hell.

Mann got ideas for his book partly from the dodecaphony (twelve-tone scale) of Schoenberg's Vienna school: 'complete integration of all musical dimensions, their mutual indifference by virtue of complete organization'. An absolute condition of the twelve-tone system, an ad hoc chosen arrangement of the chromatic twelve tones for all melodic (horizontal) and all harmonic (vertical) operations, is equal temperament. In the development of dodecaphony during the 1950s, composers went further by elaborating numerical series of such mathematical complexity that they could be adapted to all the musical parameters within a given composition. The parameters included melody, harmony, rhythm, dynamics, changes in timbre and, of course, overall form.

When the hearing processes begin to fail

Determination was total. But contrary to the complexity of some mediaeval music (mentioned above), in this very music specific results of composition would sometimes have been possible to achieve by using a stochastic working method. There often appeared, as a result of attempts to reach a complete integration of all parameters, an incongruity between a piece of music as (a) intellectual operation, semantically codified and visualized in its score, to be used as a kind of data report to check the pre-calculation relative to the guiding artistic idea (or original conception) and (b) a flow of tonal information submitted to auditory perception that aims at gestalts based on our cultural-musical codes.

Theodor W. Adorno commented often on this type of contemporary music, which had (and still has) enormous influence on the more synthetic styles followed during the 1960s and 1970s. I quote one of the comments: [3]

In the history of the mind, the relationship between the arts and sciences is not void of tension to such a degree that, in the process of progressive rationalization, the arts can turn into science and, so to speak, share in its triumph.

Here we stand at a crossroads. For the first time in the history of music, the conceptualization of musical creativity has become so complex under the influence of meta- and extra-musical ideas that auditory and associative neuronal activity—heretofore sufficient for a meaningful perception—seems to fail.

We do not know the limits, of course, to the perception of complex musical structures.* New features always need time to be effective by means of codes. Beethoven's contemporaries regarded his last quartets as very odd creations, yet their nature of 'being music' was probably never questioned; the trained ear can grasp most of these structures, even in their fine details. But some of the most important music of the 1950s was different: the possibility to grasp its structure (the emotional aura included) mainly through auditory activities—forming the central sensory mechanisms for listening—is in many cases very small. Satisfactory comprehension requires not only visual assistance to the auditory mechanisms; it needs also the possibility to go beyond the score to the prerequisite basic formulas created especially for each individual composition.

Music's genetic and evolutionary aspects

Max Weber argued that the development of Western music is characterized by an ever increasing rationalism, and he was probably

---

* *Cf.* M. Moravcsik, The Ultimate Scientific Plateau, *The Futurist*, October 1985.

right. Every step of innovation with which we have dealt in this chapter, devoted to Western traditions, means an increased conceptualization of music and, at the same time, a progressive structuralization.

The neurobiologist and Nobel laureate, Ragnar Granit, believes that when we enter the realm of music and musical creativity we must question the use of the biological concept, purpose:

There is no explanation of the talent that made possible the creation of the Ninth Symphony or the Marriage of Figaro. Why has musical creativity turned up such high levels of excellence? A possible answer is that this talent has proved harmless in the process of natural selection and so has escaped annihilation. We can, of course, supplement this with a number of postulates, such as that musicality is polygenetically determined in happy symbiosis with some more useful characteristics.

The music on which Granit, a passionate musical layman, comments belongs to the level of musical development characterized by rich innovation and manifold conceptualization; it is therefore perceived within cultural codes resting upon strong neuronal and psychological selection. But most of the world's music is not like this—it is less autonomous, more in symbiosis with activities such as rites, hunting, dance and other synergetic activity. This music is rather indifferent to the fine selective process of defining pitch, and we call such naïve music 'emmelic'.

I have suggested, in different connections, the term emmelic to denote primary melodic structures with a vast tolerance regarding pitch, intervals and durations. The term comes from the Greek *emmelēs*, in earlier Greek praxis meaning suitable in melody, or harmonious; it is used here to denote melodic patterns different from those of the spoken line of intonation but still not finally fixed and systematized as in European, Indian or other art music. Emmelic thus corresponds to a musical behaviour. It may be archaic or express a living tradition or a new direction, underlining other expressive qualities than those typical of autonomous musical behaviours as we know them (what I have called *melic*).

Musical evolution, from the time of the ancient Greeks through the Middle Ages until the situation of about thirty years ago is the story of growing deterministic thinking, the belief in a complete congruency between input and output, between what is given to the 'ear' and the perceptional result. The auditory system is thus regarded as a linear process; the frequency of a tone reaching the receptors on the basilar membrane or the auditive modules in the primary auditory cortex excites neurons that respond to exclusively defined frequencies. This is at least what we once thought.

We know today that this is not the case. In principle, every auditive neuron in the auditory cortex can respond to any frequency. There is in the auditive system a topological organization whose columns are arranged progressively as a keyboard, where the distance between columns is about one-sixth of an octave. But every column has a special arrangement of neurons, the real distance between the best responses of the columns fluctuating between 140 and 200 cents. Both the Pythagorean scale, the different temperaments, and, not least, the integrated structures of postwar music are epigenetic phenomena that—slowly through history—become codified as standards.

To sum up, the emmelic behaviour (with all its variability and absence of systematic precision) is the basic human way to make music. Emmelic behaviour is therefore always present, as well, in cultures with melic superstructures in their art music. Popular folk melodies from mediaeval times are spread all over Europe in countless melodic variants, all being adaptations to regional and local dialects. The Italian *La Folia* is represented in Sweden alone by hundreds of versions. These have all kept a link to their common origin and thus a kind of identity because of perceptual constancy.

The emmelic is, then, the ground on which any culture may build its autonomous musical systems. These musical systems will always differ in sharpness, because of the open character of the auditory field, in regard to the selected values of pitch and duration. Melic music, on the other hand, and the observance of its strongly selected and systematized parameters is probably a rather late acquisition of the human brain.

The epistemological impact of musical notation

I have tried to underline that the trends to increased musical autonomy and thus diminished symbiosis with dance, poetry and rites (such as in pagan societies, where music was a servant, a starting and moving mechanism)—trends culminating in this century—were born in Pythagorean Greece. [4]

Pre-classical Greek society (as illustrated in the Homeric narratives from the ninth century B.C. or in the famous legends, such as that of Oedipus) was a pagan, overly limbic world; it was quite different from the later Greek states of the famous philosophers who were to have such a strong influence on later European development.

It is not our task here to discuss what might have caused the radical changes in the attitudes to life between the pre-Homeric world and classical Greece. Many speculations have been made by historians, archaeologists and psychologists-neuroscientists, but I wish to refer only to Julian Jaynes. He argues that, during the period of the great city-state cultures of the Old World, there was a continuous process of change regarding the nature of consciousness: from a communal or collective one to self-consciousness as we know it now. Such a change would postulate a new and integrated hemispheric co-operation in the total working of the brain.

It is more than probable that the very diversified organization of, for instance, Sumerian society, was conditioned by a series of radical innovations regarding state economy, control of grain stores, and so on. A sort of governmental statistics was created by means of memory devices, e.g. the cuneiform linguistic signs, and codified accounting systems. All these were made possible only through an enhanced conceptualization of cognitive models, presented by 'innovators' or various kinds of experts.

This process continued through Phoenician times, when the sign system was improved by a very important step ahead. This was the change from an ideographic to a phonetic system, i.e. from a system performing the visual representation of an object or idea to a system that brought forward the language itself through its sounds. The Phoenician alphabetic system was then improved by the Greeks in two important aspects:

(1) the Greek alphabet had signs for all sounds—consonants and vowels, whereas the Phoenician syllabary was limited to signs for consonants;
(2) the Greek system of reading and writing proceeded from left to right, i.e., it was orthograde, contrary to the Arabic and Hebraic systems that kept the Phoenician retrograde, or right-to-left system. The latter is how classical Chinese, Korean and Japanese are still written.

Why a Greek input, rather than Indian?

These changes, to which I shall return in regard to their impact on the development of Western music, gave a neural congruity to the relationship between the spoken, the written and the read language. This was especially so because the production of speech, hearing and reading-writing became based on the same sequential organization and neural processing within the left hemisphere. This factor contrasted with the Sumerian/Egyptian ideographic and the Phoenician retrograde notations belonging to the right hemisphere. Thenceforth, every sign produced, heard or read derived its significance from its relationship with what went before and what came after.

It is not easy to explain why all this came to be created by the Greeks rather than by the Arabs or Jews possessing the same level of civilisation, but it certainly had an enormous impact on European conceptions of time and time experience (so different, for example, from the Indian conceptions). I believe that Julian Jaynes' views on a new consciousness emerging in the post-Homeric period of Greek history fit well in this general picture of new perceptual modes. It is through these that Greece began to leave the pagan past that it had shared with other cultures south and east of the Mediterranean Sea.

Notation has brought to musical perception and thinking a strong visual element. To follow a musical notation means reading a sequential 'text' in an orthograde direction. Physiologically, it is at this moment that music is no longer a matter for only the holistic right cerebral hemisphere. As a consequence of the visual system's double mechanism for laterality, orthograde musical reading involves the same engagement for the left cerebral half as does reading a linguistic text, i.e. use of the visual cortex in collaboration with the centre for understanding language.

The aim of a notation is, by definition, referential and conceptualizing logic. The fact that each sign in a sequential notation acquires its significance because of its relationship with the signs, coming before and after, causes (automatically) an increasing need for ever more refined definitions and qualities—but these do not necessarily have an inherent musical nature. Here a process of formalization begins with regard to proportions, relations and values.

Music as cultural counterweight

Strong conceptualizations together with autonomous musical thinking, perhaps caused or accelerated by the existence of a sequentially organized notation, are the real background to Max Weber's notation of the rational character of Western music. Music is not language in the strict biological sense, yet language has played an enormous role in Western music: not as an object for singing activities nor because of possible common roots with music, but as a mechanism used by the cognitive processes.

I suppose that it would be reasonable to argue that a society with a high degree of division of labour, and possessing developed rational technologies that diversify and separate people into groups of varying tasks and interest, needs mental activities to counterbalance an increasing lack of communal fellowship and strongly fixed common values. Society needs, in other words, mental activities corresponding to the level of rational and technological thinking on the one hand, and simultaneously corresponding to the human need for limbic expression on the other.

This is what European music has accomplished during its history; it has increased musical structuration to ever more complicated patterns corresponding to the increased complexity of society (science, general knowledge of the physical world, economic structure, and so on) but, at the same time, in accordance with growing mental complexity resulting from the growing complexity of human ecology. The exchange of information and values between music and society is different than in a basically emmelic society—but the processes are the same.

In the case of Europe, this has led to a situation (culminating during the past two centuries via their revolutionary, political, social and scientific development) in which music, in its art forms (in contrast with African societies, with their more synergetic life modes), represents autonomous values. Sometimes we speak of 'absolute music'. The conception of such music in the mind of the innovator—the specialized composer—can be triggered by the most varied general ideas and experiences; in the creative process, these are then transformed into musical structures.

The triggering idea is probably always accompanied by a kind of emotional aura serving as its driving force. The task of the composer is to translate the idea, with its aura, into musical structures by means of the technical devices of composition. This is a question of how to move from general impressions and intentions to increasingly clear and concrete musical formulations. (The *musical* material usually comes more into the focus of the composer's attention.) The material of this transformational process, however, is not only of musical character. Also in action is one's inner web of thoughts, emotions, associations and earlier experiences.

The aims of music as a cognitive system

It is difficult to pass judgement regarding the biological purpose of music in the complex human context. Where the 'melic', or strongly conceptualized and sophisticated, music reaches a summit, there may exist a biological purpose in the form of values, values expressed as exquisitely structured *Gestalt* and emotion. The purpose is one thing, however, and the way the purpose is genetically based is another.

The auditory system may have its origins in the 'lateral line' system of certain fishes. This system is one of balance and orientation. A reason for the development of the lateral line via an amphibian (e.g. the frog) into the acoustical orientation organ of terrestrial animals is the need to localize and identify countless ambient sounds. When a certain level of evolution had been reached, the powers of speech and language developed from basic organs whose primary functions were breathing and the ingestion of food. Complementary adaptation of the brain's function and structure may have been the radical response to how the total organism coped with challenges in an ever more complicated world.

The anatomical asymmetry of the parts of the left hemisphere responsible for speech, language perception, writing and counting suggests a more direct genetic background of the speech functions than that valid for music. (An anatomically specified correlate to musical ability has not been found in either hemisphere.) Functionally, however, the repetitive and acoustically relatively stable sung, blown or stroked tone seems generally to activate neuronal fields in the right hemisphere. This is not because this hemisphere is 'musical', but because its kind of intelligence favours an appositional and holistic behaviour in response to stimuli, and those stimuli spontaneously creating *Gestalt* are at home here. On the other hand, whether this background could indicate—at emmelic level—a genetic state constituting a specific ability is doubtful. The melic level is so mixed and complex that Granit's hypothesis of a polygenetic determination seems probable. Melic music is probably not a product of nature as is speech. It is an artefact, what Stravinsky described as 'nothing else than a phenomenon created by mental activity'. It may be possible to argue that this music is the most human of all creations, or that the study of musical structures within the perspective of new theories such as Ilya Prigogine's notion of dissipative structures—or René Thom's catastrophe theory—may open new doors to science.

Music and mathematics

I hope that it is clear from what I have said that melic music—regardless of its European, Indian or Chinese cultural roots, and irrespective of its grade of conceptualization—is not equivalent to science. Music has played a very central role, in all these cultures, in the development of the notion of science. But the role and purpose of music are different from those of the sciences. Music is a cognitive modus bridging holistic and discursive views of life and the world; it helps us to enrich the rational mind with the necessary and modulating colours; music helps us to control or to give a counterbalancing perspective to our limbic mind, with its strong undercurrents.

It could be true, however, that some aspects of the creative processes in music and science—especially mathematics—come very close and that some common views on our world may unite them, from an epistemological point of view.

The ties between mathematics and music have been said to be very close, and the examples given above do not contradict this view. New biological, especially neuroscientific, knowledge does not support the thesis, however, without further qualification. Counting, for example, is an ability chiefly of the left cerebral hemisphere, while tonal sequences are at home in the right hemisphere. There is a qualification to this: the more conceptualized the music, the more the hemispheres co-operate. A practical result is that the professional musician, the specialist, is more of a 'left-hemispheric' listener than the naïve listener. Nonetheless, music on the whole is an activity characterized by the right hemisphere's ability to recognize entities or *Gestalten*. Let us say, therefore, that if music and mathematics meet, the meeting place may lie in geometry instead of arithmetic.

Oswald Spengler once prophesied, in speaking of the science of the recent Western era, that it should 'carry all the traits of the great art of counterpoint'; he represented the 'infinitesimal music of outer space' as our culture's 'deepest longing'. Perhaps. In any case, this returns us to our original query.

Does the sun 'ring and peal'?

If our new knowledge seems headed in a direction where we shall be able to speak about an inner counting not so much in purely mathematical or physical but in biological (especially microbiological) terms, then is there—on the other hand—correspondingly new knowledge at the macro-level?

It is indeed possible to argue along these lines. We know that the universe is filled by the most diversified kinds of sound resulting from celestial events of the past, a distant past whose relationship with Pythagoras' experiments with the monochords is like a comparison between infinity and less than a second ago. ∎

Notes

1. They considered music *exercitum arithmeticae occultum nescientis se numerare animi* (the secret arithmetic exercise of the unconscious mind).

2. Polyphony refers to musical composition characterized by parallel
   (simultaneous) and harmonizing, yet melodically autonomous and individual,
   parts or voices.
3. *In der Geschichte des Geistes ist das Verhältnis von Kunst und Wissenschaft
   nicht derart spannungslos dass bei fortschreitender Rationalisierung jene zu
   dieser würde und gewissermassen an ihrem Triumph teilhütte.*
4. From my own paper, Privacy of Sound, read at a symposium with the same
   title arranged by the Council of Europe in Cork (Ireland), 14-16 May 1985
   (publ. No. 3.624).

To delve more deeply

BOULEZ, Pierre: 'The history of music is not only the history of feelings and
   musical thinking, but also the evolution of material sound.' See also Chapter
   24 of the present volume.
GRANIT, R. *The Purposive Brain*. Cambridge, Mass., MIT Press, 1977.
HANDSCHIN, J. Die Musikanschauung des Johannes Scotus (Erigena),
   Deutsche Vierteljahrschrift fur Litteraturwissenschaft und Geistesgeschichte,
   Vol. V, 1927.
——. *Der Toncharacter. Eine Einführung in die Tonpsychologie.* Zurich,
   Atlantis Verlag, 1948.
JAYNES, J. *The Origin of Consciousness in the Breakdown of the Bicameral
   Mind.* Boston, Houghton Mifflin, 1976.
SEIDE, Stuart. As quoted by T.Q. Curtiss in the *International Herald-Tribune*
   of 28 February 1987, this contemporary American professor of mathematics-
   turned-dramatic director and actor relates the science of numerology to the
   theatre: 'In both there is an element of illusion. . .as unanswerable as the
   axioms of geometry.'
STRAVINSKI, I. *Poétique musicale*. Cambridge, Mass., Harvard University
   Press, 1947.
WALLIN, N. Biological Aspects of the Relationships between Music and
   Languages, *Diogenes*, No. 122, 1983.
——. Pitch Perception as Expression for Exogene and Endogene Coordinated
   Oscillations, *The World of Music*, No. 3, 1983.
——. Further Remarks on a Biological Approach to Musicology, *Trends and
   Perspectives in Musicology* (Proc. World Music Conference, International
   Music Council, 3-5 October 1983; publ. No. 48;. Stockholm, Royal Swedish
   Academy of Music, 1985.
——. Reflections on What is Called Absolute (Perfect) Pitch, *Analytical Studies
   in the Description and Analysis of Music in Honour of Ingmar Bengtsson,
   2 March 1985* (publ. No. 47), Stockholm, Royal Swedish Academy of Music,
   1985.

An art historian describes the cultural significance of the development of linear perspective in biplanar reproductions of reality, how and why this technique emerged concomitantly with the Italian Renaissance, and its consequences for modern Western, Muslim and Asian civilizations and economies. Perspectival evolution directly affected knowledge and achievement in architecture, engineering, medicine, the handicrafts and economics, as well as the fine arts themselves.

Chapter 10

The origins of objective representation in art and science

Samuel Y. Edgerton, Jr.

Amos Lawrence Professor of Art History at Williams College, Dr. Edgerton is the author of The Renaissance Rediscovery of Linear Perspective (1975). *The essay comprising the present chapter is based on* 'Linear Perspective and the Western Mind: The Origins of Objective Representation in Art and Science,' (Cultures, Vol. III, No. 3, 1976), *since extensively revised and rewritten. The author's address is Graduate Program in the History of Art, Williams College, Box 8, Williamstown, MA 01267 (United States).*

The early beginnings of linear perspective

Why did the West rise so suddenly to dominate the rest of the world after the sixteenth century? Why did China and Islam, two civilisations ahead of Europe in technology and science during most of the Middle Ages, drop so precipitously behind after the European Renaissance? These are questions which have vexed historians, philosophers, and even psychologists and politicians for years. The problem has been raised again by the distinguished scholar Joseph Needham in his ambitious project, *Science and Civilisation in China,*[1] as he details in page after page the remarkable inventiveness of the mediaeval Chinese while Westerners were still wallowing in their 'dark ages'. Why did not this precocious activity of China (and also Islam) immediately translate itself into a genuinely modern age of scientific discovery?

In thumbing through Needham's vast work and pondering the wealth of indigenous illustrations he has used to prove the early superiority of Chinese technology and science, one is struck by something that has up to now escaped even his own attention. This is the fact that none of the Chinese pictures is drawn in linear perspective. Not one of the marvelously complex scientific and mechanical devices has been represented in the way we take for granted today as realistic (Figure 1). Receding, parallel edges of surfaces, as depicted by the Chinese even as late as the twentieth century, are never shown as converging, the way we think such lines should always appear in simulated reality. Distant things are also never represented as relatively smaller than those nearer, such as makes us recognize depth in photographs and Western Renaissance-style paintings.

How strange! For all their scientific precocity, the Chinese (and Muslims) never in their early heyday realized the 'scientific truth' of linear perspective. One wonders how Chinese craftsmen could have fashioned full-size, operating models from such 'naive' pictures. Indeed, it comes to mind that linear perspective was a unique idea of the West. No artist in any other civilisation ever arrived at this idea independently without the direct or indirect influence of the Italian Renaissance. Furthermore, linear perspective came about at the very moment when the West began to pull ahead of China and Islam in matters of science and technology. While Needham did take note of the lack of perspective in Chinese art in general, he offers no correlation between that and the Chinese failure to keep up with Western science after the sixteenth century.[2]

Is it possible that a mere matter of art—that luxury we are able to indulge in only after life's real survival issues have been resolved—can

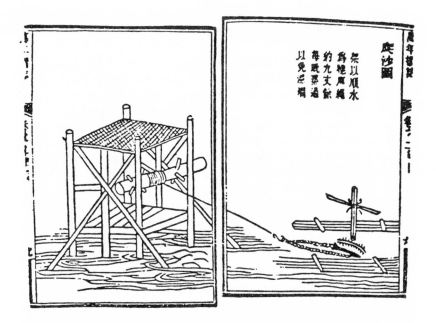

Fig. 1. Illustration of a dredger, from Wan-Nien Chiao Chih *(Record of the Bridge of Ten Thousand Years) by Hsieh Kan-Thang (Nanchheng, Chiangsi Province, 1896).*

have played a crucial role in what 'really counts' in the modern scientific and technological world? As an art historian, I have become convinced that just as life begets art, so can art beget life. In fact, I should like now to propose that the advent of linear perspective in the West did have a profound effect upon the 'real' world. It not only allowed such 'reality' to be depicted, it created a new psychological mastery of that reality. It prepared the West to launch its own age of technological and scientific discovery by the late fifteenth century. Conversely, the lack of perspective in Chinese and Islamic art may have been a key factor in inhibiting these civilizations from competing effectively with the West until they themselves adopted the Western artistic style.

Non-perspectival conventions in East and West

Let us go back in time to the moment when China and Islam were still ahead of the Christian West in the quality of their respective science and technology. Let us choose a date of about 1300 and pick at random three contemporaneous pictures, one each from China, Islam, and western

Europe. Let us look, for example, at a Chinese depiction from *Nung Shu*, a popular illustrated treatise on agricultural technology, or an Arab illustration of a similar subject from one of the many beautifully illuminated *al Jazari* manuscripts, or any Christian miniature of the time like those in artisan handbooks following the tradition of Villard d'Honnecourt. As we examine these separate pictures, we observe, in spite of obvious regional stylistic differences, a remarkable parallel in the way all three artists have tried to render an illusion of the third dimension. None of them knew linear perspective, yet each used almost identical conventions for showing how solid objects position themselves in space. Furniture items such as tables are always shown at angle, with the most important side—the top—exaggerated and distorted. Background buildings are flattened out, often with more sides showing than could logically be seen from a single viewpoint. All three artists took for granted that their viewers would automatically recognize objects at the bottom of their pictures as nearer and objects at the top as farther away, even though depicted as the same size.

What are we to make of such coincidences? Dare we say that they reflect some shared outlook on life as well as art? Could there have been such a thing in 1300 as an all-prevailing mental set common to all peoples of the world at the time, regardless of their cultural differences, and which is now quite foreign to our modern visual psychology? At least we can state that these pre-perspectival pictorial conventions have something to do with a vaguely understood structuring process in human intelligence. Somewhere in the human brain there exists an innate relationship between what is seen in the phenomenal world and certain basic forms for representing it in pictures. This relationship does *not* have to do with linear perspective, however. Similar conventions are also found in all pre-Renaissance art from the cave paintings of the Cro-Magnon men, the tomb paintings of the Egyptians, to the pictures of so-called primitive peoples today. We see it also in the art of little children, before they become perspectivally acculturated as adults.

The odd thing about linear perspective, to repeat, is that it is not innate. It must always be learned. Even though modern science claims perspective approximates geometrically they way light rays are projected physically into the eyes, there is no evidence of any *a priori* structure in the mind which makes possible an intuitive transfer of the perspective image to a picture. Peoples from cultures outside the Western tradition often have trouble at first 'seeing' perspective pictures. Even black and white photographs, invented to reproduce mechanically the phenomenon of Renaissance perspective, are not instantly recognized as realistic by certain 'primitive' peoples.

The illusionary 'window' in the West

Why then did Christian civilisation after 1300 abandon such a universally understandable system of pictorial communication for difficult-to-learn linear perspective? Why did this happen only in the West? Let us look now at three more pictures, one each from China, Islam, and the West, this time from about two centuries later; for instance, a Ming Dynasty decorated scroll by a great artist like Shen Chou, or an illuminated page from one of the many beautiful Persian *Shāh Nāmah* manuscripts like that in the Houghton Library at Harvard University, or an Italian Renaissance painting by a master such as Raphael. In the first two we would notice little difference from anything earlier. The artists of China and Islam are still not interested in improving 'realism' by tinkering with their traditional tried-and-true pictorial conventions for representing spatial illusion. However, this is decidedly not the case in the Western picture. What we observe here is that Raphael (as well as all western European painters by this time) is now employing a whole new vocabulary of pictorial conventions for representing the third dimension. He often deliberately chooses bird's-eye or worm's-eye perspectives that demand novel arrangement of volumes in space in order to remain consistent and optically 'correct.'

But the fundamental difference between the Western example and those of the East at this time is not so much in the new pictorial conventions as in the whole conception of what a picture is. The Western artist has begun to think of his picture, not as a decorated surface, but as a 'window'—a hole in the wall—through which the viewer is to image himself seeing the actual event. This is a notion that does not concern the Chinese or Muslim painter. While Chinese idioms for depicting volumes in space were relatively convincing (even by modern perspective standards), the Chinese artist always understood his picture as an integral part of the same flat surface on which he inscribed his beautiful calligraphy. The painting is a caesura of contemplation between literary messages, never an isolated, visual hole in the wall. Western artists, on the other hand, began to segregate their pictures from any accompanying text. If literary inscriptions were included, they frequently appear on banners or other objects within the picture and are part of the same overall illusion. This new window conception was to make demands on both artist and viewer that had no parallel in China or Islam.

Remembering one's own first picture

Perhaps a personal analogy might give some hint of the profundity of this change, not just of the mechanics of making pictures but of the very psychology of seeing itself. Recall, if you can, that moment long ago, in nursery school or kindergarten, when you painted your first picture.[3] There was that big, empty white paper on the easel and all those brimming, beckoning pots of poster paint! It is not clear what you thought a picture was in those days, but it wasn't a 'window' (in fact, the window picture was another exercise altogether, when the teacher asked you to trace on a pane of glass what you saw on the other side). Anyway, you were literally enwrapped in that first painting on paper. It was a surrounding space you were colouring, not a see-through space at all, a surrogate world; not a photographic representation, where you could pour out your fantasies and insecurities. A blue streak at the top sufficed for sky because everybody knows you see sky when you look up. The opposite for ground; a green streak (providing you live in the suburbs) was enough to show that. Your house you delighted to show with all sides at once . . . of course, because you knew it had four walls since you played around them all the time.

What a shock it was therefore when you arrived in primary school, and, as you learned to read and write, you found yourself under the influence of a new kind of pictorial world, that of photographs and perspectival, textbook illustrations. Your old standard of pictorial realism, once based so confidently on personal intuition, no longer obtained. Whatever talent for drawing you once had now suffered ignominious put-downs. Maybe your teacher still encouraged you, but your friends . . . Their withering comments ('What's *that* supposed to be?') made you think it was time to move on to something else for peer approval. Whether you realized it then or not, you were undergoing a psychological transformation so complete you could never 'see', let alone depict your intuitive world again.

The above analogy suffers of course because it confuses the climactic changes in Western Renaissance art with the biological growth of a child, implying thereby that Chinese and Islamic art—and their concomitant ability to 'see'—remained 'childish' as the West 'matured'. Indeed, there are some thinkers who have expressed this view, such as the psychologist Jean Piaget ('phylogenesis recapitulates ontogenesis').[4] Such a notion hardly does justice, however, to the tremendous, continuing sophistication of Chinese and Islamic art forms, which managed to achieve the most stunning subjective expression even if they did not lend themselves as well to 'objective truth'. In the nineteenth century, these continuing ideals of the East, through the medium of the Japanese print, returned again to nudge Western painting once more from its window obsession and back to decorated surface.

Images of art and reality of vision

My own interpretation of children's art, however, is that it is no mere deficient form of adult expression. It is a wholly separate and self-contained system of depicting and is not necessarily improved by perspective. In this sense it can be compared to Chinese and Islamic art. Just as in children's art, the adult, sophisticated Eastern artist did not consider the blank spaces around his painted objects as having volume. Pictorial space was only the neutral flatness of the paper. The relation of the depicted objects to this space and to each other was determined only for design reasons that take into consideration the overall shape of the paper. Just as little children seem to see their own imaginative world in terms of this kind of pictorialization, so it may be true that even adults in historic civilisations before the advent of linear perspective saw nature in a different way than we do today. The art historian, Millard Meiss, has documented how the fourteenth-century St. Catherine of Siena described a miraculous vision in which Christ and the holy saints appeared before her 'just as she had seen them painted in churches'. In other words, the reality of her vision was corroborated only when she could match its appearance with images in contemporary art![5]

The Western artist of about 1500, however, had clearly abandoned this traditional pre-perspectival attitude. By thinking of his painting as a window, he had learned to understand space as volumetric rather than flat. It had to envelop his depicted objects almost like some transparent fluid. Therefore, the artist had to depict the blank space around objects as if it were an object in its own right and subject to the same geometric structure.[6]

The Graeco-Roman precedent

Our problem is not to judge the aesthetic quality of art, with or without the window option. We are interested only in why the West adopted it and what impact of that notion was upon the *Weltanschaung* of the West vis-a-vis the East. One basic reason why the West could adopt this notion, we shall now take up, was because there existed a precedent already in classical antiquity. To some extent, Chinese art and Islamic art were also influenced by Greece and Rome, but Western Christianity of course received the classical tradition most directly.

Observe, for example, how many Roman wall paintings from the still-preserved houses in Pompeii are decorated with imaginary picture frames, surrounding scenes clearly meant to be understood as if they were happening outside, beyond the walls themselves. Perhaps I am sounding redundant as I keep stressing this window concept, but the fact is it was unique to Graeco-Roman civilisation. Never before had it occurred, in the art of Egypt or elsewhere. By late antiquity, this conception almost led to the invention of linear perspective.

The rise of Christianity, with its dependence on Eastern ideas, aborted any further movement in the perspective direction. Nevertheless, in spite of the long hiatus, from the fall of the Roman Empire until about 1250, even as Western Christians exploited the same pictorial forms as the Chinese and Muslims, there remained a vestige of Graeco-Roman tradition in Western art. We need not resort to Jungian archetypes to demonstrate this. Erwin Panofsky has convincingly shown how conventions of Graeco-Roman proto-perspective remained in a kind of suspended animation in mediaeval Western art, as if waiting to be revived again by application of Renaissance 'warmth and moisture'. [7]

Geometric optics in antiquity

But why, in the first place, did the Graeco-Roman artist resort to the idea of a picture as a window? At this point, we may turn to the history of science for an answer, once more coming up with something unique in Western civilization. This was the science of geometric optics, or how the eye sees according to Euclidean rules. Euclid, Ptolemy and other Greek mathematicians had determined that light always travels in straight lines, therefore light rays could be diagrammed on a piece of paper.

It followed that the way images are formed and the way they appear distorted because of position or distance can be worked out precisely by the geometric laws which govern cones, pyramids, and triangles in mathematics. The parallel science of optics in China, incidentally, was never concerned with such a geometric explanation for vision; Euclidean geometry itself did not arrive in China until the seventeenth century.

Greek physiologists, like Galen, added further knowledge to the science by the time of the Roman Empire. Their basic tenet was that the visual cone did not simply terminate at the eye but within it, bringing a small-scale image to the sensitive membrane according to the same Euclidean rules that allow a small triangle to have exactly the same proportions as a larger, similar one. Galen understood, as did all optical scientists until the sixteenth century, that this sensitive membrane, on which the small-scale image was displayed within the eye, was what we now call the lens. In the Latin of his day it was called the *crystallinus* and was understood as a transparent body, like a window, intersecting the visual cone before the rays finally converge at their apex on the optic nerve. We now know that the lens is only a focussing device, directing the incoming light upon the retina, the true seat of vision within the eye. However, the notion of the intersecting *crystallinus*, even though erroneous, was to have important effects upon not only the way Graeco-Roman painters conceived of their pictures, but on the rebirth at the Renaissance of linear perspective.

The force of Islamic influence

We have no specific evidence that classical artists actually applied these optical ideas to their art. We have some vague references in Vitruvius and Lucretius that stage designers, at least, were creating optical effects with their scenery. We also know that Ptolemy, in his writings on geography, had invented a method of cartographic projection that comes very close to Renaissance linear perspective. We shall be discussing the importance of this further on, but for the moment, we can only say that classical artists seemed to be paralleling in their pictures what contemporary scientists were explaining as essential to the visual process itself: that the eye sees the same way a window 'receives' images upon it as an intersection of the visual cone.

Oddly for our story, it was Islam that first rescued classical optics from the chaos of Rome's collapse. Muslim scientists like Alhazen and Avicenna carried it to even higher development than had Euclid, Ptolemy or Galen. Nevertheless, there was little interest in Islam for bringing the science to the attention of Muslim artists. This was because of Muslim religious iconoclasm, discouraging any kind of figurative expression in the arts as tampering with God's handiwork. Certainly, iconoclasm did not prohibit figurative art everywhere in Islam, but it served to limit the use of illusionistic expression as one of the didactic tools of the Muslim faith. In fact, oriental attitudes toward pictorial art were originally similar to those which had come about in the West only since the nineteenth century, i.e. that they should appeal to man's subjective nature and not attempt to teach moral lessons about religion, politics or science.

The revival of optics and its relation to theology, society and art

After the retreat of Islam from Spain and Sicily in the twelfth century, a great residue of scientific books by Aristotle, Euclid, Ptolemy, Galen and others, with rich Arab commentary, were left behind. These books looked so intriguing that a veritable army of Christian scribes converged to copy them into Latin. Among the most sought after were texts on optics. Why, we might wonder, would such a mathematical science find appeal in the Christian West? For an answer, we need only turn to the writings of Roger Bacon, Franciscan monk of the late 1200s and author himself of a treatise on optics and of a most interesting polemic on the obligations of painters.

For Bacon and the other Christian commentators of the Middle Ages, geometric optics held a singular fascination. Since God had created light on the first day of Genesis, and since the science of optics explained how light travels according to immutable geometric laws, it would seem therefore that optics contained the proper model explaining how God spreads His divine grace throughout the universe. Here is a relevant passage from the *Opus Majus* (1267), Bacon's great compendium of all knowledge containing also a section on optics:

Since the infusion of grace is very clearly illustrated through the multiplication of light, it is in every way expedient that through the corporeal multiplication of light there should be manifested to us the properties of grace in the good, and the rejection of it in the wicked. For in the perfectly good the infusion of grace is compared to light shining directly and perpendicularly . . . since they do not reflect from them grace nor do they refract it from the straight course which extends along the road of perfection in life . . . .But sinners, who are in mortal sin, reflect and repel from the grace of God 8

Few religions in world history have been as dogmatic as Christianity, that is to say, so blatant in attempting to subsume every aspect of human experience and knowledge under the moral dictatorship of the Church. The Christian fathers felt compelled to co-opt every new scientific thought that came along. All aspects of knowledge, they believed, had to be rationalized according to the Revealed Truth of Scripture and fit precisely into the moralized structure of 'natural law'. It was very much in this vein that the science of optics came to be appreciated again in the West during the Middle Ages.

The guardians of orthodoxy in China and Islam, on the other hand, more or less tolerated both science and art because these pursuits usually did not conflict with the basic order of their societies. Artists and scientists in the East also often enjoyed an elitist status, and this too may have lulled them into being a little less inquisitive than their more insecure Western counterparts. While art and science, both West and East, were subject to local political winds, in the West they fell especially under the intense scrutiny of an international church. Some new ideas, like advancing the nude in art or the heliocentric theory in science, might win temporary favour but eventually had to be condemned by the ecclesiastical authorities. Other new notions, like increased geometrical severity in art or the concept of infinite space in science, might be ignored at first but then be championed by the same church. Success for the Western artist and scientist was therefore more than a mere matter of temporal fame. It had very much to do with his eternal salvation or damnation. Indeed, the course of art and science in western Europe during the Middle Ages and Renaissance provides historic proof that sometimes, in human progress, 'malign attention' can produce just as much creativity as 'benign neglect'.

The beginnings of middle-class capitalism

In this vein again, let us read Roger Bacon's admonition to artists, remembering that it was written about 1260, at the time when his own Franciscan mother church at Assisi was preparing to receive that revolutionary group of painters, led perhaps by Giotto, who would turn the course of Western art irrevocably towards the rediscovery of linear perspective. Bacon himself was hardly an art enthusiast. He appreciated artists only insofar as they met their responsibility to rehearse the didactic message of the Church. He also had no foreknowledge of perspective, yet he clearly understood the moral implications of an art based on the laws of geometry:

Now I wish to present the . . . [purpose] . . . which concerns geometrical forms as regards lines, angles, and figures both of solids and surfaces. For it is impossible for the spiritual sense to be known without a knowledge of the literal sense. But the literal sense cannot be known unless a man knows the significations of the terms and the properties of the things signified. For in them there is the profundity of the literal sense, and from them is drawn the depth of spiritual meaning by means of fitting adaptations and similitudes, just as the sacred writers teach, and as is evident from the nature of Scripture, and thus have all the sages of antiquity handled the Scripture. Since, therefore, artificial works, like the ark of Noah, and the Temple of Solomon and of Ezechiel and of Ezra and other things of this kind almost without number are placed in Scripture, it is not possible for the literal sense to be known, unless a man have these works depicted in his sense, but more so when they are pictured in their physical forms; and thus have the sacred writers and sages of old employed pictures and various figures, that the literal truth might be evident to the eye, and as a consequence the spiritual truth might also. For in Aaron's vestments were described the world and the great deeds of the fathers. I have seen Aaron thus drawn with his vestments. But no one would be able to plan and arrange a representation of bodies of this kind, unless he were well acquainted with the books of the

Elementa of Euclid . . . and of other geometricians. For owing to ignorance of these authors on the part of theologians they are deceived in matters of greatest importance . . . . Oh, how the ineffable beauty of the divine wisdom would shine and infinite benefit would overflow, if these matters relating to geometry contained in Scripture, should be placed before our eyes in their physical forms! For thus the evil of the world would be destroyed by a deluge of grace . . . Surely the mere vision perceptible to our senses would be beautiful, but more beautiful since we should see in our presence the form of our truth, but most beautiful since aroused by the instruments of vision we should rejoice in contemplating the spiritual and literal meaning of Scripture because our knowledge of all things are now complete in the Church of God, which the bodies themselves sensible to our eyes would exhibit. Therefore I count nothing more fitting for a man diligent in the study of God's wisdom than the exhibition of geometrical forms of this kind before the eyes. Oh, that the Lord may command that these things be done! 9

Bacon's conception of the role of art would have been quite foreign to China or Islam. He not only envisioned an art totally at the service of the church but one also ordered by the same mathematical laws with which God orchestrated the universe. Bacon, in fact, was writing at the very moment when we begin to trace the spread of bourgeois capitalism in Western Europe. It was the time of 'secularizing the sacral, and sacralizing the secular', as Charles Trinkaus has expressed it.

Make no mistake, however. The appearance of materialistic capitalism in the West did not at first cause any inverse decrease in religious zeal. Quite the contrary. Every effort was made to keep God and Mammon reconciled. As Bacon himself urged, the church must co-opt even the most materialistic ideas in order to defeat Islam and convert the world. The church's best weapon in fact was the science of mathematics. If mathematical laws are the model for God's orchestration of the universe, then it follows that any application of these laws in the secular world must bring God's grace down to earth.

The permeation of wholly new attitudes in the West

No wonder then that the fourteenth century saw such a burst of mathematical activity in all fields of human endeavour. People entered into such studies with the belief that God granted his providence, both here and in heaven, to those whose affairs were most mathematically ordered. Much of this effort, interestingly enough, began in Italy, especially around the city of Florence. Here at the time was invented the double-entry system of bookkeeping, which not only improved accounting but lessened the risk of the entrepreneur as he gambled his capital for higher and higher stakes. New methods for charting the trade routes on the Mediterranean Sea were devised. The mechanical clock appeared, making possible an orderly work day no longer dependent on the seasonal sun. New systems of land distribution (the *mezzadria**) came about, with crops dispersed in geometrically arrange terraces, increasing their yield. One could go on and on. Nowhere in China or Islam at this time was there such a systematic dependence on mathematics both to solve man's material problems and, simultaneously, to have these efforts blessed in heaven as morally in tune with God's will. One is reminded of the late-fourteenth century Tuscan businessman, Francesco di Marco Datini from Prato, who closed his ledgers each night with the inscription at the top of the page, 'In the name of God and of profit'.

Thus there evolved in Italy during the late thirteenth century an attitude that was to permeate all Western art within the next three hundred years. This was quite at odds with what was happening contemporaneously in China and Islam. The leaders of Western art, from Giotto to Raphael, considered painting at its best when expressing the moral lessons of the church, its purpose achieved only when the composition was visibly geometric. More and more, pictures came to be understood in Europe as instruments of instruction rather than of poetic contemplation. This was to have fundamental importance for the study of science, as we shall see.

Western re-discovery of linear perspective

Brunelleschi

It was in this didactic, geometrizing vein that Western art was led inexorably to linear perspective. The actual moment when the precise rules were worked out, derived as they were from optics and the artistic incentives of Giotto and his followers, was as late as 1425. The individual who put these ideas together was the Florentine sculptor and architect, Filippo Brunelleschi. I shall not go into the details here since I have already described them in my book *The Renaissance Rediscovery of Linear Perspective.* [6] Instead, let us look again at our imagined painting by the Italian painter Raphael; this time we'll identify an actual work by the master, the *Betrothal of the Virgin* or *Lo Sposalizio*, originally painted in 1504 and now in the Brera Gallery in Milan.

Here is an excellent example showing Brunelleschi's fundamental discovery of how the 'vanishing point' works in linear perspective.

* The *mezzadria*, a system now defunct, provided for sharecropping between the tenant and owner of farmland.

Sometime around 1425, eighty or so years before Raphael's picture, Brunelleschi had noticed, perhaps by looking in a mirror, that the phenomenon we all experience as we view rooms or roads apparently diminishing in distance is explainable because their receding edges seem to come together at a point or points always on the same level as the viewers' eyes. This horizontal level we now call the horizon line. When drawn on a picture, the horizon line indicates both where the painted figures should be placed and where the viewer should position his own eyes in order to recapitulate exactly what the artist had originally seen.

Hence, the horizon line establishes an optical bond connecting the viewer's real space and the picture space beyond the fictive pictorial window. Furthermore Brunelleschi realized, if the artist/viewer stands foresquare in the center of a quadrangular room (as by implication in Raphael's painting), the receding edges will seem to converge to a single vanishing point opposite the eyes. It is as if an imaginary visual ray were passing between, at once perpendicular to the surface of the eyes and the surface of the picture. Such a centric visual ray (Figure 2), the only one in the visual cone perpendicular to both the *crystallinus* and the surface of the object seen, had long been recognized by the classical, Arab and Christian optical scientists as crucial for carrying the visual image most clearly and distinctly to the intellect. All the other visual rays enter the eye obliquely, hence they bring in the image less clearly.

Raising the nobility of the visual arts

As Roger Bacon had observed, oblique light rays cause refraction and reflection and can be likened to the way God's grace is repulsed from sinners. The centric ray, on the other hand, is not reflected or refracted, hence it defines mathematically how God's grace descends to the good. Therefore, when Brunelleschi discovered how the vanishing point works, he made it possible for artists not only to picture a scene realistically, but also metaphorically, in the sense that the vanishing point covers the position in the painting that connects most clearly and distinctly with the viewer's very soul. Here then was the place in the picture to depict the most meaningful element of the scene. Leonardo da Vinci, for instance, painted the head of Christ directly over the vanishing point in his Milan *Last Supper*.

In Raphael's *Betrothal of the Virgin* the vanishing point is in the open doorway of the circular building in the background. While this may seem an innocuous place, it had religious significance for the Renaissance viewer. The building in fact represents the mystical temple of the church, the open door of which, with obvious symbolic ramifications, connects

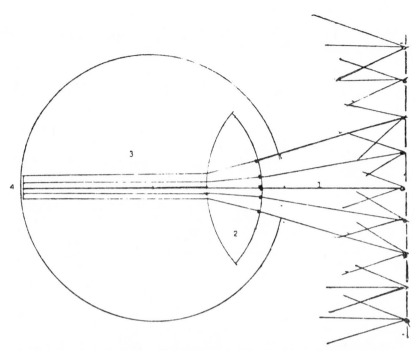

Fig. 2. Diagram of the eye as understood in classical, Arab, and mediaeval Christian optics; it shows Alhazen's idea of how the visual rays (1) enter the pupil, implant their 'form' on the crystallinus (2) and are then refracted in the vitreous humor (3) in order to bear the image upright to the optic nerve (4).

via the centric ray with the viewer's eye. The door is shown open all the way through the building, so that the sky beyond is visible. Raphael has thus neatly tied the geometric notion of the vanishing point to the mystical idea of God's infinity in heavens, perhaps the ultimate realization of Roger Bacon's admonition to make literal the spiritual sense. Interestingly, Raphael's contemporary and fellow native of Urbino, Luca Pacioli, was writing on a number of relevant subjects at the same time that Raphael was painting, on double-entry bookkeeping, the five regular solids in geometry, and on the relationship between fine calligraphy and 'divine proportion' in mathematics. He also remarked that linear perspective raised the visual arts to a more noble level than music, since they now pleased the eye, which, more surely than the sense of hearing, leads directly to the soul. [10]

But surely the most consequential of the new pictorial conventions devised by the perspective painters of the Renaissance was the 'grid'. We see it prominently in Raphael's *Betrothal of the Virgin*, as the device organizing the middle ground before the circular building and behind the principal figures. Its perpendicular lines are supposed to represent a horizontal, squared pavement, the receding edges of which converge uniformly on the single vanishing point in the building's doorway. Such a grid pictorial convention, modern psychologists aver, stimulates an innate response in human intelligence, engendering a remarkable illusion of a flat surface receding into distance.

Linear perspective and cartography

The peculiar power of the converging grid had been observed by Western artists long before Brunelleschi. It was almost a cliche among the proto-perspective painters in Italy and France during the fourteenth century, but was hardly ever exploited by the artists of China or Islam. What is of particular significance about its perspective redefinition in the fifteenth century is that it came into vogue at exactly the time and place of another cognate idea, this time in the science of cartography. This had to do with the longitude-latitude system for drawing the surface of the world, originally detailed in Ptolemy's *Cosmographia*, a classical treatise that strangely escaped Western attention in the twelfth century. It arrived in western Europe for the first time only in 1400, via Florence, where it may well have been examined by Brunelleschi himself.

What Ptolemy described in this treatise were three different methods for solving an age-old cartographers' problem—how to draw an image of a curved surface on flat paper. The Alexandrian author proposed to divide the universe, which he understood as a sphere, into a system of abstract enveloping lines, vertical longitudes converging at the poles and crossed by horizontal latitudes in non-convergent but parallel rings. Since he was an optical scientist as well as geographer, Ptolemy urged his would-be viewers to assume a frontal position before a model of a gridded globe representing the cosmos and fix his centric visual ray upon a position just above the equator, the center of the world as known in antiquity (Figure 3). The squares on the world's surface near where the centric ray strikes are then seen in their normal size, but those on the northern and southern extremities, where the longitudes curve around towards the poles, appear foreshortened. Ptolemy was quite aware of this apparent optical distortion, and so he tried to devise a map technique that would let the viewer 'correct' the illusionistic difference in distance between the longitudes and latitudes on his map. [11]

Once more, this sudden appearance in the West of a cartographic system based on the abstract division of the earth's surface into geometric squares had been long preceded in the East. Islam, in fact, had managed to preserve Ptolemy's *Cosmographia* amongst all its other

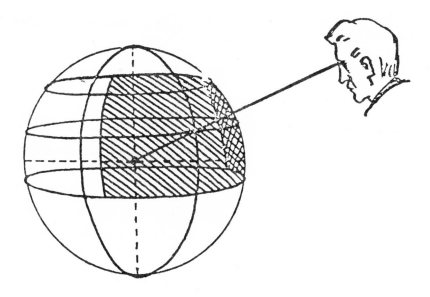

Fig. 3. Diagram showing how Ptolemy would have the viewer look at the globe, fixing his visual axis on a point in the centre of the space to be mapped (the oikumene), *just above the equator.*

treasures of classical learning and, just as with optics, had embellished it with rich new commentary. China, while not knowing of Ptolemy directly, had nonetheless worked out independently a system of cartographic representation based on the same visual psychology of the grid. Neither civilisation, however, established a relation between this cartographic technique and art.

Paolo dal Pozzo Toscanelli

The relationship between Ptolemaic cartography and art in the West began to develop almost immediately after its arrival in Florence. One of Brunelleschi's personal friends was Paolo dal Pozzo Toscanelli, a man much involved in the study and dissemination of Ptolemy's geographic theories. This friendship is attested to in contemporaneous accounts of Brunelleschi's life, as is also the mutual interest both maintained in the science of mathematics. There is good circumstantial evidence that Brunelleschi, himself without a university education, consulted Toscanelli on the matter of optics while he was working out the first linear perspective experiments. Circumstantially too, there is good reason to believe that Toscanelli, a university-educated medical doctor with extensive background in geometry, optics, astronomy, and geography, showed Brunelleschi the cartographic projection systems of Ptolemy, putting the 'bee in Brunelleschi's bonnet' to solve once and for all how the grid gives the illusion of perspective recession.

Leon Battista Alberti

Ten years after Brunelleschi's climactic rediscovery of linear perspective, another Florentine humanist and even closer friend to Toscanelli, named Leon Battista Alberti, wrote a famous book, *On Painting* (1435-1436), in which he described for the first time verbally how to draw a perspective picture. Alberti's system is clearly derived from Brunelleschi, whom he acknowledged. He also specifically demonstrated how to draw the perspective grid, as a *pavimento* (he used that word). Alberti also gave advice on how the artist should first look upon the scene he was about to paint. This is of utmost importance to our thesis. What he advised was that the artist make a *velo* of strings tied together forming a grid. This he should hang in a frame, 'so that the visual pyramid passes through . . .' (Figure 4). A similar, scaled grid should then be drawn on the artist's panel. The advantage of this:

'. . . is that the position of the outlines and the boundaries of the surfaces can easily be established accurately on the painted panel; for just as you see the forehead in one parallel, the nose in the next . . ., the chin in one below . . . so you can situate precisely all the features on the panel . . . which you have similarly divided . . . you can [also] see any object that is round or in relief, represented on the flat surface of the veil.' [12]

Alberti's statement is almost a paraphrase of similar instructions on map-making by Ptolemy. Not only was Alberti a friend of Toscanelli, he also designed Ptolemaic-style maps himself, in particular one of Rome showing the locations of classical monuments. [13]

Alberti's writings move us beyond, however, mere technical methods for drawing. His treatise on painting was a manifesto of a new way of seeing the world as well as depicting it. The reader (more often a rich patron than any of the painters to whom he rhetorically dedicated his book) was admonished to look at nature not as a mass of disorganized details but rather as a homogeneous, mathematically organized abstraction. The whole must be comprehended before the parts. Today, when we make the compliment, 'You have a good perspective of the

*Fig. 4. Illustration of an artist drawing from a 'veil' (velo). From Johann II of
Bavaria and Hieronymus Rodler,* Ein schön nützlich Büchlein und Unterweisung
der Kunst des Messens (*A Fine and Useful Booklet and Instruction on the Art of
Measuring*), *Simmern (Germany), 1531.*

problem . . .', we mean exactly what Alberti meant in his stress on linear
perspective—not just to make an image look natural, but to design it as
if nature herself were as rational as mathematical law. Alberti was very
concerned that the subject matter of painting always be worthy of a
didactic, ennobling message to the viewer. He was also sure that if these
subjects were depicted according to the rules of linear perspective, the
viewer would be encouraged to see with the same orderly rationale.
Alberti's attitude toward rational seeing is the same as that which we
expect today from scientists when they record their experiments.

An advisory to the Portuguese monarch

It cannot be proven that the Chinese or Muslims in the fifteenth century were conversely deficient in these 'seeing' activities. Nevertheless, there was not in their societies at the time the same obsession with mathematical organization that beset the West. Not only artists and map-makers were addicted to grid-thinking, but architects, city planners, military engineers, merchants, politicians, and even priests in the pulpit.

It is unfortunate that we know so little of Toscanelli. As a humanist and university scholar, he was probably even closer to Alberti than to the lower-class artisan Brunelleschi. How exciting and far-reaching must have been the conversations between Alberti and Toscanelli, about their shared interests in mathematics and cartography! One may imagine these two long-robed scholars also discussing matters of art, how linear perspective works in a painting, and how the gridded *pavimento* is a useful device for measuring the extent of land surface, both on a map and in a picture. Toscanelli was certainly familiar with many paintings showing checkerboard-like floors according to linear perspective, which had already been painted in his life-time, when he wrote this remarkable letter—perhaps one of the most decisive in history—to an aide of the king of Portugal:

Paul the physician to Fernam Martins, canon of Lisbon, greetings . . . . On another occasion I spoke with you about a shorter sea route to the lands of spices than that which you take for Guinea. And now [your] Most Serene King requests of me some statement, or preferably a graphic sketch, whereby that route might become understandable and comprehensible, even to men of slight education. Although I know this can be shown in a spherical form like that of the earth, I have nevertheless decided, in order to gain clarity and save trouble, to represent [that route] in the manner that charts of navigation do. Accordingly, I am sending His Majesty a chart done with my own hands in which are designated your shores and islands from which you should begin to sail westward, and the lands you should touch at and how much you should deviate from the pole or from the equator and after what distance, that is, after how many miles, you should reach the most fertile lands of all spices and gems, and you must not be surprised that I call the regions in which spices are found "western", although they are usually called "eastern", for those who sail in the other hemisphere always find these regions in the west . . . . The straight lines, therefore, drawn vertically in the chart, indicate distance from east to west, but those drawn horizontally indicate the spaces from south to north . . . . From the city of Lisbon westward in a straight line to the very noble and splendid city of Quinsay (China) 26 spaces are indicated on the chart, each of which covers 250 miles . . . . So there is not a great space to be traversed over unknown waters. More details should, perhaps, be set forth with greater clarity, but the diligent reader will be able from this to infer the rest for himself. Farewell, dearest friend.[14]

The letter is dated 1474, eighteen full years before Columbus's voyage of proof. Toscanelli's map, sad to say, no longer exists but it, along with the letter, were passed on to the Genoese sailor. Indeed, Columbus kept up a correspondence with the ageing Florentine doctor; many historians believe that Toscanelli's encouragement was the major factor in holding Columbus steadfast to his project until the Spanish finally backed it in 1492.

Why was America 'discovered' first by the West? Was not Chinese and Islamic nautical technology every bit as good as the West's, as witness the maritime expansion of these two powers in the Indian Ocean during the fifteenth century? Why was there then no Chinese or Muslim Columbus?

Perhaps, we might ponder, because there was no Chinese or Muslim Toscanelli. Nor, apparently, could there be, because, as we have seen, there lacked completely in China and Islam the kind of dialogue between artists and scientists which had become endemic in the West since the late thirteenth century.

Linear perspective and the organisation of space

Today we are the tired children of the Renaissance. We are too used to linear perspective. Instead of being excited by its novelty as were the Europeans of the fifteenth and sixteenth centuries, we easily become bored with looking at so many repetitiously 'realistic' pictures in our art museums. However, if we could only abandon our modern preferences for a more subjective and decorative art, we might again appreciate how much that Renaissance perspective tool beckoned people to a deeper contemplation than ever about the nature of man and God. In particular, linear perspective in pictures seemed to encourage the Renaissance viewer to think about the *space* of the universe not as something finite and heterogeneous, but as a continuous quantity, uniform, infinite, and homogeneous. I believe that this initial Renaissance exhilaration upon looking at perspective pictures was not unlike that which a person, who has been used only to walking, feels during his first experience in an aircraft. Local obstructions and deviations that concern and deter the pedestrian suddenly disappear and are of no consequence when comprehended from the air.

The effect of perspective pictures and Ptolemaic maps must have been similar. May we consider that Christopher Columbus was moved by such pictures? The traditional, mediaeval fear of the mysterious ocean, for instance, that had kept mankind landlocked for so long, began to give way under the power of the perspective grid. We can imagine Columbus arguing with his detractors as to how the great ocean should be seen, not as a series of unaccountable, unrelated emotional confrontations, but (as Toscanelli had imagined it) a mere extension of Ptolemaic squares, mathematically abstracted and therefore subject to man's rational will. We should not forget, incidentally, that Christopher Columbus was born less than two hundred kilometres from Leonardo da Vinci and was his exact contemporary.

Linear perspective and scientific representation

Leonardo da Vinci

For the same reasons we have been discussing, there was no counterpart at this time in China or Islam to Leonardo da Vinci. Great artists and great scientists certainly abounded in the East, but they did not come together in one person. People are still mystified by the genius of Leonardo, regarding him as some kind of genetic mutant. Perhaps it would be better, however, to comprehend him as a very keen mind born in the right place at the right time. Perhaps the luck of Leonardo was not so much his high I.Q. as his proximity to Florence at the very moment when Alberti's and Toscanelli's ideas were percolating down to artisan practice. To the credit of the Creator, furthermore, Leonardo was endowed with twin ability in art and science unequaled in any other contemporary. In spite of his more renowned achievement in painting, Leonardo remained by his own reckoning primarily a scientist. That is to say, he always wanted to be a great engineer and inventor, carrying out vast projects in fortifications, armaments, water works, etc., for prestigious patrons.

For Leonardo, linear perspective was also no mere recording convention. It was integral to the very way he observed nature. By 'seeing' in perspective, Leonardo could grasp, better than any experimental scientist before him, the patterns and structures that underlie physical and biological phenomena.[15] He was certainly aware of perspective power when he wrote, 'Painting . . . compels the mind of the painter to transform itself into the very mind of nature, to become an interpreter between nature and art. It explains the causes of nature's manifestations as compelled by her laws.'[16]

Galileo

Galileo Galilei, the greatest scientist of the seventeenth century, was, as we by now should expect of a native Florentine, a skilled artist with special talent for linear perspective. In 1609, he constructed a telescope after the recent Flemish invention, and focussed its manifying power on the moon. While not the only scientist at the time to do so, he was certainly the first to recognize the lunar surface for what it really is, a pock-marked landscape with lofty mountains and concave craters, as he himself represented it in his epochal book, *Sidereus nuncius* (Starry Messenger), published in Venice in 1610 (Figure 5).

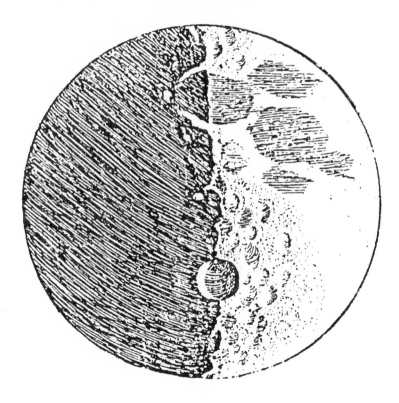

Fig. 5. An illustration dating from 1610 that was one of the first to show the moon's features as three-dimensional characteristics. The satellite is shown here at first quarter (from Galileo Galilei, Sidereus nuncius, *Venice, 1610).*

Why was Galileo able to make this observation of the moon's topography when no one else had ever seen such a condition, even through a similar telescope? Because Galileo already knew what a sphere with an irregular surface should look like *before* he looked through his own instrument. He had, in fact, practiced drawing such figures in his study of perspective: for instance, how a sphere with raised protuberances under raking light would be shaded and shadowed, as in Figure 6, an illustration Galileo would certainly have known from Daniel Barbaro's *La practica della prospettiva* (Venice, 1568), the most popular text on the subject in Italy.

Oddly enough, perspective books like Barbaro's were not yet published or circulated in Britain, where the mathematician Thomas Harriot had already observed the moon through a telescope a few months before Galileo. Harriot's eyesight, unfortunately, was as yet untrained by exposure to perspective drawing. He could only see the magnified moon according to the tenets of his mediaeval Aristotelian education, as a perfectly smooth ball superficially mottled by an unexplainable 'strange spottednesse'. Harriot had no inkling that the odd discolouring of the moon was caused by hollows and protuberances illuminated by bright sunlight on one side and casting dark shadows on the other, as Galileo recognized instantly from his earlier familiarity with Barbaro's perspective problems! [17]

Vesalius

So far we have been discussing only the remarkable relationship between perspective and the mathematical/physical sciences. It also had just as profound effect upon the biological sciences, particularly medical anatomy. If we were to compare again some sixteenth-century scientific illustrations from either Chinese or Arabic treatises on the subject with any picture in Andreas Vesalius' *De humani corporis fabrica* (On the Fabric of the Human Body) published in Basel, 1543 (Figure 7), we would quickly note how the former show, as usual, the organs and muscles of the human body depicted in impressionistic, flat linear designs. There is no sense of depth; there is no way to tell by looking at such illustrations which organ is in front or which behind or how it works. In fact, the pictures are useless if one accepts that their accompanying texts were intended to teach students where to find the depicted organs in actual human bodies. Needless to say, the Vesalian picture is revolutionary. Even modern photographs of the same muscles and organs are not as clear as Vesalius' illustrations. Never before Vesalius' time could the student have at his or her disposal such a learning tool.

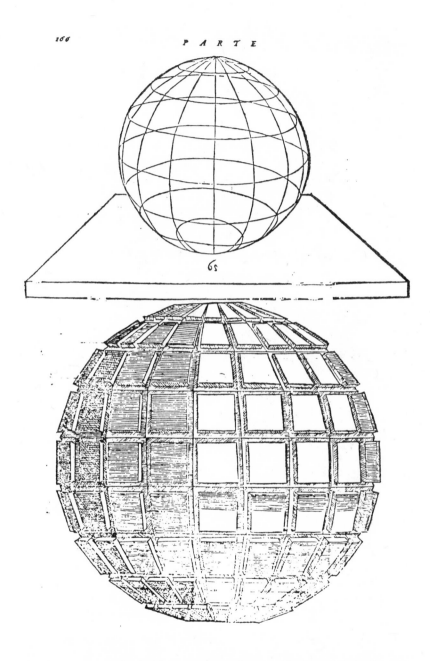

Fig. 6. A sphere is divided by vertical meridians and horizontal latitudes (top), then given perspective by adding protuberances seen under raking light (bottom), from Daniel Barbaro, La pratica della prospettiva, *Venice, 1568.*

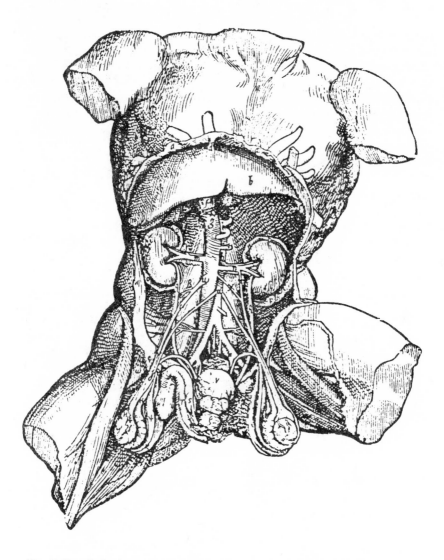

Fig. 7. Revolutionary anatomical drawing of the torso (1543) by Andreas Vesalius gives depth to size and position of skin, muscles, organs and other components, from De humani corporis fabrica, *published in Basel.*

It should be pointed out that Vesalius was only the most brilliant of many anatomy-book writers in the sixteenth century. A number of such illustrated treatises began to be used in the medical schools of Europe (at first no better in teaching how to cure than the schools of the East at the same time). These truly revolutionized the training of doctors.

Until Vesalius' time, the medical schools of Europe pedantically depended on the ancient and often erroneous textual descriptions of the structure and physiology of the human body according to Galen. Because of certain religious and social taboos, dissections of cadavers were few and far between. Not only was it difficult for students to study the body empirically, but, worse, there was no effective way for recording what was revealed during these rare occasions. As any 'pre-med' student knows who has ever had to suffer through a practical exam, there is no substitute for a good perspective illustration. No amount of words, not even photographs, can replace it.

Significance of these advances in perspective

To be sure, Vesalius and his colleagues were not directly dependent on linear perspective for the superiority of their anatomy illustrations. Nonetheless, Renaissance concern for the human nude was a major by-product of such geometric fascination. Perspective drawing first made it possible to represent the human figure in all sorts of complex poses, no longer restricted (as in mediaeval art) to frontal or profile views. This new ability aroused interest, moreover, in the contraposite (antithetical) views of classical Greek and Roman sculpture. As artists became more used to the picture as a window, they were intrigued to depict the human figure as antique statues. (See Figure 8.)

Vesalius and his colleagues cleverly took advantage of this increasingly popular notion. They borrowed flagrantly from currently admired art, stripping it of skin, and exposing the muscles and viscera as if these manikins of *situs* were stylish sculpture—note the 'broken' limbs in Figure 7—like the Belvedere *Hercules* torso, a statue much copied by Raphael and Michelangelo.

Earlier artists, like Leonardo and Michelangelo, paved the way for Vesalius by applying their own newly trained perspective vision to the minute study of the human body both externally and internally. Since the late fifteenth century, artists, not doctors, possessed the most knowledge about human anatomy. Until Vesalius the subject was virtually an artists' monopoly, and to this day few believe Vesalius could have executed the illustrations himself. They are conventionally attributed to a hack painter named Kalkar but, more likely, they are by the great Titian himself.

Linear perspective and the scientific revolution

The rediscovery of linear perspective preceded by only a few years the European improvement of the printing press. Once more this was a

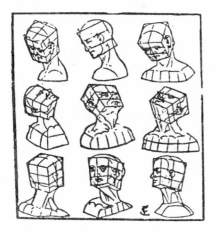 

Fig. 8. The human figure became reproducible in a variety of complex poses, thanks to perspectival drawing. These foreshortened heads and bodies are from Erhard Schön, Unterweisung der Proporzion (*Instruction in Proportions*) *of 1538.*

fortuitous coincidence, considering that China had had printing already for centuries. However, perspective pictures and the printed word together then proceeded to revolutionize mass communications.

One cannot overstress the importance of the printed, *perspectivally illustrated* textbook to the increasing power of the West in all matters of science and technology after the sixteenth century. Now it was possible for more people than ever before to learn not just a wide variety of subjects, but to see the world according to uniform geometric laws which, up to a point, encouraged an accurate mechanical understanding of how all physical objects, microscopic as well as macroscopic, relate to one another as volumes in space. Through the mass availability of perspective pictures, anyone could, theoretically at least, be a Leonardo da Vinci.

No wonder then, that within the span of hardly a hundred years, the West was able to produce a Galileo, a Kepler, a William Harvey (who learned his medicine at Vesalius's University of Padua), a Descartes, a Newton, and a Linnaeus. I would be much too presumptuous to say that these great geniuses of the Western scientific revolution owed their contributions directly to linear perspective. It is impossible to imagine, however, how they would have done what they did in a world without the technique.

A state of mind about the world

This chapter began on an upswing note about the West; it now ends with a Spenglerian conclusion. At the very moment that the German philosopher was penning his dire predictions, Western artists were happily giving up their didactic, linear perspective tradition in favour of a flat and decorative style we now call modern art. Paradoxically, and also at the same time, Russia, one of the last Western nations to experience the Italian Renaissance, and one of the first to experiment with modern art at the turn of this century, decided, under its Soviet leadership, to re-embark upon its interrupted Renaissance perspective course (now 'socialist realism') in tandem with its crash programme of modernization.

Japan and China too, in spite of the singular artistic heritage of those Eastern cultures, saw fit to adopt Western Renaissance-style perspective for textbook illustration and public art—doing so even after Western artists had begun to show their admiration for Eastern painting.[18] Odd it is that social scientists studying the Third World today rarely notice how often industrial modernization and artistic perspectivization are symbiotically related in many non-Western cultures. Indeed, they need only examine the political posters and postage-stamp designs from these 'developing' countries. Some even accuse the Western powers of 'cultural imperialism,' i.e. of deliberately suppressing non-Western indigenous art forms.

But linear perspective, as I have tried to show, is not merely an artistic convention like decorative painting. It is, rather, a state of mind—a way of structuring not a picture but the visual world. Perspective is psychologically related, for better or worse, to scientific imagination and technological literacy. If, as some believe, the West is destined to decline as the East gains in technological capability, Westerners can at least take comfort in the fact that, whatever culture is to overtake them, its people will first have learned to 'see' according to the linear perspective of the Renaissance. ■

Notes

1. J. Needham, *et al., Science and Civilisation in China*, Cambridge, Cambridge University Press, 1954-.
2. Chinese artists traditionally rendered the illusion of pictorial depth by means of oblique parallel lines, sometimes called axionometric or isometric perspective. See Needham, *op. cit.*, Vol. IV, Pt. 3, Sections 28-9, p. 104. For

a discussion of similar but less systematic conventionality in Islamic art, see Donald F. Hill, *The Book of Knowledge of Ingenious Mechanical Devices*, Reidel, Dordrecht-Boston, 1974. Both Needham and Hill have pointed out rightly that pictures rendered in such a manner are more advantageous to technical drawing since, theoretically, all depicted dimensions remain undistorted. (In fact, most modern engineering drawing is of this sort.) I would reply, however, that neither Chinese nor the mediaeval Muslim artists intended their drawings to be used as scale models for constructions—as Western designers did after the fifteenth century. Scaled axionometric drawing also came about first in the West, as a by-product of linear perspective. This allowed engineers and architects an advantage: they could thenceforth conceive and work out, entirely on paper, complicated three-dimensional structures. The ability to build from scale plans is thus another, uniquely Western, invention.

3. If one needs reminding, see Rhoda Kellogg, *Analyzing Children's Art*, Palo Alto, Stanford University Press, 1969.

4. See Herbert Ginsburg, Sylvia Opper, *Piaget's Theory of Child Development: An Introduction*, Englewood Cliffs, Prentice-Hall, 1969, p. 207.

5. Millard Meiss, *Painting in Florence and Siena after the Black Death*, Princeton, Princeton University Press, 1951, p. 105.

6. On the notion of 'space' and the various ways it has been signified in Western art, see Erwin Panofsky, Die Perspective als symbolische Form, *Vorträge der Bibliothek Warburg, 1924-25*, Leipzig, 1927, p. 258; on the changing ideas of 'space' in science, see Max Jammer, *Concepts of Space: the History of the Theories of Space in Physics*, Cambridge, Mass., Harvard University Press, 1969. Also relevant are S.Y. Edgerton, Jr., *The Renaissance Discovery of Linear Perspective*, New York, Basic Books, 1975; Fritz Winckel, Space, Music and Architecture, *Cultures*, Vol. I, 1974, p. 135.

7. Erwin Panofsky, *Renaissance and Renascences in Western Art*, Stockholm, 1960.

8. Robert Belle Burke (ed., trans.), *The Opus Majus of Roger Bacon* (Part VI), Philadelphia, University of Pennsylvania Press, 1928, Vol. II, p. 238.

9. *Idem.*, Vol. I, p. 232.

10. Luca Pacioli, *De divina proportione*, Venice, 1509; a German translation with commentary is Constantin Winterberg (ed.), *Luca Pacioli: Divina Proportione*, Vienna, Quellenschrift für Kunstgeschichte (Neue Folge), Vol. 2, 1889.

11. Concerning Ptolemy's *Cosmographia* and its mapping methods, see Edgerton, *op. cit.*, Chapters VII and VIII.

12. Cecil Grayson (ed., trans.), *Leon Battista Alberti, 'On Painting' and 'On Sculpture*,' London, Phaidon, 1972, p. 68.

13. See Luigi Vagnetti, La 'Descriptio Urbis Romae' di Leon Battista Alberti, *Quaderno n. 1, Università degli Studi di Genova, Facoltà di Architettura*, October 1968, p. 25.

14. Samuel Eliot Morison, *Journals and Other Documents on the Life and Voyages of Christopher Columbus*, New York, Oxford University Press, 1972, p. 11.

15. This is also the thesis of Kenneth D. Keele, whose *Leonardo da Vinci's Elements of the Science of Man* (London, Academic Press, 1983) argues that linear-perspective geometry formed the basis of Leonardo's theory of natural law.
16. Philip McMahon (ed., trans.), *Treatise on Painting by Leonard da Vinci*, Princeton, Princeton University Press, 1956, Vol. I, p. 41.
17. For a detailed discussion of Galileo's and Harriot's lunar observations in the context of Renaissance perspective theory, see S.Y. Edgerton, Jr., Galileo, Florentine *disegno*, and the 'strange spottednesse' of the moon, *Art Journal,* Vol. XLIV, 1985, p. 225.
18. For an account of how Chinese artists tried to adapt their traditional style to the copying of illustrations in European scientific and technical treatises brought to Asia by the Jesuits in the seventeenth century, see S.Y. Edgerton, Jr., The Renaissance Artist as Quantified, In: Margaret A. Hagen (ed.), *The Perception of Pictures*, New York, Academic Press, 1980, Vol. I, p. 179.

See also:

D. Hockney, Vogue par David Hockney, *Vogue* (French edition only), December 1985/January 1986, pp. 219-259. Here Hockney, according to Ruth Marshall of *Newsweek*, seeks to 'eliminate the distance between the spectator and the painting by giving [biplanar] lines a forward momentum, which draws the viewer into the work of art'. Eds. Condé-Nast, 4 place du Palais Bourbon, 75007 Paris (France).

Martin Kemp, God's House, Geometry's Handiwork, A Renaissance Formula for Rendering Church Interiors, *The Sciences*, January/February 1986. Kemp is professor of fine arts at the University of St. Andrews.

Creation consists of motivation, the collection of germane data, and then the synthesis and evaluation of all available information. The creative process may originate in the unconscious, although not in a vacuum. It invokes memory, the cultivation of awareness, cumulative knowledge or habit, self-organisation, and often intensive and even 'possessed' labour. Also vital are introspection, vision, dreams and the capacity to distinguish the peripheral from the essential.

Chapter 11

Beyond the looking-glass

Gérard Blanc

The author is a graduate of France's Ecole Polytechnique and the University of California at Berkeley. A consultant on science and technology policies and on technology assessment, Mr. Blanc is co-founder and editor-in-chief of the French quarterly journal, CoEvolution, *and he has taught management courses in France and Switzerland. He wishes to thank Professor Lynn Margulis for her 'invaluable help in preparing this essay in English.' The author is currently technical consultant to M2I, 11 Bis rue de Balzac, 75008 Paris (France). Tel: +331 4289-0809.*

The response from the mirror to him
  who gazes into it, to the gazing stranger, is a man.
The response from the mirror to Wholeness
  is part of the Whole.
Its response to the universal is the specific.
Its response to the possible is the fact, the object.
Its response to substance is the accident.

                                        Paul Valéry
                                        Narcissus (1926)

Every experience of creativity possesses a mysterious, ineffable aspect that escapes from language—no vocabulary adequately and completely approaches it. To capture the essence of creativity, creators use metaphors. In replacing infinitely varying ideas and actions with an ordinary word, a single metaphor suffices.

Does the metaphor of the mirror open new windows on creativity? In a figurative sense, the mirror provides mankind with the means to know itself and ways of knowing anything by producing the mirror's image. The idea of the mirror suggests the look, reflective vision, right/left symmetry, differences between two nearly identical images—all fundamental to the creative process. The mirror evokes direct observation, the observation of one's behaviour, interaction between different parts of one's being. But the myth of the mirror reminds us too of Narcissus' sterile self-preoccupation, of his dying—ever incapable of transcending his own image.

This essay deals with individual creativity (regardless of field) seen as a psycho-physiological process: self-communication. The creator co-evolves within his individual context and peculiar natural and social environment. The metaphor invokes the following actions: to look, observe oneself, perceive resemblances, integrate differences, go beyond the reflection. These will be examined in the light of recent ideas.

Creativity, function and process

Creativity, as used here, is meant in its broadest sense: in the arts, science, technics, human relations and family life, professional and political life, and so on. In this sense, everyone is a creator (in one way or another), but often he or she is the only one to know this. Sometimes even the creator is not aware of his act. Thus, the unity of the creative spirit is considered, beyond its multiple forms, to be a property of all living beings.

Our hypothesis begins from this concept, as David Abram has recognized.[1] For Arthur Koestler, creativity is not specifically human; it is discernible at each level in the evolutionary hierarchy—from the simplest unicellular organism to the genius of humans. Creativity therefore becomes the realization of abilities revealed in abnormal or exceptional circumstances that translate into novel behaviour. At work at each of these stages are, according to Koestler, 'homologous principles.'

If creative action is considered a function of living beings, we are then led to ask 'biological questions': What is its selective advantage? How is creativity transmitted? How has creativity become more complex through evolution?

From the point of view of geological time, sexuality is an early manifestation of the creative processes. That sexuality generates creativity has long been known to the poets, but more recently it has been studied by biologists. One can consider both birth and creativity as processes and functions: both resemble, and respond to, the challenges posed by environmental change—whether physical or organisational, whether natural or mental.

The study of creativity, notably in technical fields or business management, is usually limited to the solving of problems clearly posed, whereas creative behaviour usually extends over a far vaster domain. A problem is rarely formulated in terms that are easily resolved, so a task of the creator consists precisely of defining the problem and identifying the constituents of an acceptable solution—both of which are discovered 'along the way'.[2] Novelty, remaining as an idea in the mind of its inventor, is not creation; it must be translated into communicable form—a genetic program, or another system of comprehensible symbols.

Communicating creativity

Edward de Bono has provided a definition of the problem, embodying these different aspects: 'A problem is simply the difference between what one is and what one wishes to be.'[3] This includes all the cases of creativity that do not respond directly to the exigencies of the external environment but to necessity. Internal tensions within the creator make him unable to do anything, as it were, but create in order to liberate himself. From such a vantage point, problem solving and creativity are the same.

One does not study the creative process from the outside. So the metaphor of the mirror finds here its first application: Many creators observe themselves, relating their experiences one after the other, forming a puzzle from which emerge patterns. Life histories of innovators do not suffice. Books dealing with the psychology of invention, drawing on well-documented famous works, distract us from everyday manifestations of the creative function. The loving looks of a mother to her infant, appeasing words that stop a nascent quarrel—these are related to the same general creative process that characterizes masterpieces of art or great inventions.

In science and mathematics (with some exceptions, e.g., Poincaré or Schrodinger), very few researchers reveal the way in which they were led to their discoveries. Dissimulation of the steps anterior to the formal presentation of proof finds its culmination among mathematicians of the formalist school (the Bourbaki group, for example): they refuse to reveal the gropings of their predecessors.

Poets willingly speak of their creative work, but only by communicating via poetry itself. Thus a Poe or Verlaine or Eluard neither analyzes nor teaches.

Patterns of the creative process emerge when one combines the observations of a great number of creators and examines them through different, contemporary points of view: through biology, neuro-psychology, information and communication sciences, theories of self-organization, 'transpersonal' psychology, and so on.

Vision

Life does not exist in the absence of exchange of material and information from the environment. The evolution of life shows increasing complexity of the sense organs. Among the senses, vision plays an essential role; in western civilisation, it seems to become the dominating sense (see Edgerton, Chapter 9). Paul Valéry has noted that the 'consciousness of animals seems to appear with the organs of vision that permit the evaluation of a situation *as a whole*'.[4] One must not neglect the other senses, however; the Soviet mathematician Pontryagin and the Argentinian writer Borges bear witness to the fact that blindness does not impair creativity. I use the term 'vision' to include all forms of sensory perception.

In this sense, images that nourish and put creativity into action constitute, at the same time, its starting ingredients. Contemporary neurophysiology joins Buddhist teachings: vision is not passive. It spirits and actively participates, even if unconsciously. Spontaneous activity of the nervous system itself produces images in the waking state as in sleep.

The brain acts as a filter; it selects facts furnished by the environment and constructs reality as it is perceived. Koestler has described this distortion, inherent in our psychophysiological mechanisms:

man always looks at nature through coloured glasses . . . the 'innocent eye'
is a fiction, based on the absurd notion that what we perceive in the present
can be isolated in the mind from the influence of past experience. There is
no perception of 'pure form', but meaning seeps in and settles the image.[5]

Mathematicians and poets suggest that creativity begins, indeed, with
vision: 'intuition is the faculty that teaches us to see' (Poincaré);
'invention is only a way of seeing' (Valéry).

But there is vision, and there is vision. The painters at the beginning of
the 20th century emphasized to their benefit the relativity of vision. Paul
Klee observed that

what we see is a proposition, a possibility, an ersatz. The real truth is first of
all invisible. If formerly things were represented just as they were seen on
earth, things which one loved or would have loved to see, today the relativity
of the visible has become obvious.[6]

Cubists change space, propose objects that, without losing anything of
their external appearance, are enriched by internal perspectives, as if the
eye had the power to wander within. Through art, man becomes
conscious of an image that is distinguished from the common myth. Art
destroys cliché.

To learn to be a creator is to learn to begin to see, to see differently
and simultaneously from different angles. Gregory Bateson has shown
the importance of double description, such as that of binocular vision:

from this elaborate arrangement, two sorts of advantage accrue. The seer is
able to improve resolution at edges and contrasts . . . More important,
information about depth is created. In more formal language, the *difference*
between the information provided by the one retina and that provided by the
other is itself information of a *different logical type*. From this new sort of
information, the seer adds an extra *dimension* to seeing.[7]

A broader vision

To discover is to uncover. To reveal is to show, expose, make known.
Creativity involves the removal of covers that conceal, the exposure of the
hidden. Valéry has said, 'the solution existed and nothing was changed,
but one had only to see it—previously it was not recognized as the
solution'. Koestler, commenting on the discovery of Archimedes'
principle (according to legend), has told us that everyone entering a
bathtub knows that the water-level rises. (We all check to see that the

tub will not overflow.) Archimedes knew this, too. It must have been his pressing concern, the need to measure the density of the King of Syracuse's crown, that made him suddenly conscious of the displacement of the water. The ordinary was perceived from a new point of view. Koestler showed that this applies as much to scientists as to artists, 'who make us see familiar objects and events in a strange, new, revealing light. Newton's apple and Cézanne's apple are discoveries more closely related than they seem'.

To discover is also to unveil implicit principles, to query the bases of a theory, of attitudes, of ways of action. The inventor is a critic, acting as a spoilsport in his irreverence. Einstein denied the existence of the ether and reference to the absolute in order to describe the motion of bodies. The twelve-tone musicians contested the principle of tonality, the hierarchy of notes and intervals. Cézanne replaced detailed sketches, including those of faces, with simple volumes.

Actions outside the mainstream

Our problems, our impasses, often emerge from *idées fixes*, from rigid concepts. That which the Anglo-Americans call insight (note that its etymology emphasizes the role of vision)—intuition, essential perception—requires, according to the physicist David Bohm, that one abandon one's presuppositions about the observer and the observed, about one's ego, about the nature of psychological time, and so forth.[8]

It is not surprising that change in specialization or discipline stimulates scientific research. René Dubos, trained initially as an agronomist, had a series of successful careers as agricultural engineer, microbiologist, epidemiologist, ecologist, economist and writer. Only imperfectly trained as a microbiologist (always an outsider in the discipline), Dubos discovered gramicidin, the first antibiotic to be produced commercially.*

Creations and inventions are apparently rare in the 'mainstream'; they seem to be made 'on the side'. The most fertile ideas are often those least expected, those for which there is little formalized research, that appear when the creator considers the absurd or abnormal or bizarre, when he asks himself 'Why not?' and tries to challenge the conditioned responses.

In order to nurture creativity, Edward de Bono teaches 'lateral thinking' whereby he invites us to see situations from as many different angles as possible. He advises us not to move one's thoughts simply

---

* Gramicidin is a crystalline polypeptide produced by a soil bacterium, active against bacteria producing disease in local infections.

forward, but to invoke many directions. Based on movement and change of perspective, lateral thinking makes jumps, moves through uncertain stages, integrates fortuitous intrusions, and explores improbabilities.

Recombinations

Human creativity emerges from new combinations born of pre-existing ideas. Creators do not begin from nothing; they combine elements according to a complex and ubiquitous process of recombination. Koestler has spoken about 'collisions of frames of reference', reminding us that the Latin *cogito* (I know) means literally, 'I shake together'. Einstein, in one of his letters to the mathematician Jacques Hadamard, explained his discovery of relativity by a 'combinatory interplay between reproducible mental elements'. Henri Poincaré noted that

> to create consists precisely in not making useless combinations, but in making that very small minority of useful ones. Invention is discrimination, is choice.[9]

This assertion, frequent among numerous creators, still does not have a clear, scientific base. The analogy with sexual reproduction, however, may clarify the situation. Sexual organisms result from new combinations of well-adapted genes. By recombining different characters in a single individual, hybrid vigour is achieved: living beings increase their chances of overcoming the challenges posed by the environment.

Continuing the metaphor, the selection of ideas can be envisaged in a similar manner. Talking about selection presupposes mechanisms of choice. With respect to human creativity, the creator appears as subject, equipped with intentions and motivations, in both the individual and collective contexts. In order to explore the creative context, the creator must observe himself—as the mirror invites him to do.

Looking inwards

The critical research on artificial intelligence conducted by Herbert Dreyfus shows clearly that people function differently than machines. His work teaches us the manner in which choices between multiple combinations are made, as discussed above. Dreyfus has sanely reminded us that the world of man is prestructured, that objects are given meaning only by the preoccupations of the moment.[10] This inability of machines to distinguish the essential from the accessory impedes definitively the possibilities of artificial intelligence. The computer's behaviour cannot be guided by emotions or personal worries.

Furthermore, affective pressures such as the death of a spouse, adulation of a military leader, the beauty of a flower, a thunderstorm's actions on the senses, a long and painful illness, religious conviction, an encounter with a charismatic person, hunger, sleepiness—all these are always expressed within the body. Such corporal imperatives furnish, directly or indirectly, a direction for action to be undertaken. Computers lack bodies, and will always lack them.

Necessities and intentions are not all conscious. The least known and least comprehensible aspects of the creative process arise from the unconscious, but this does not mean from a vacuum. All creators emphasize the prerequisite of preparation, of laborious infusions identifying the creator with what will become of his creation. This symbiosis, this deep association (like that between mother and embryo) seems to be a required stage in the creative process. The creator becomes literally possessed by his work. Thus the ideas of Henri Bergson and Einstein, developing from problems that plagued them both as adolescents, were converted into significant bodies of knowledge. Bergson continued to be concerned with the distinction between time and duration; Einstein needed to know how the concept of time could be communicated.

Analyzing the surging of technical design—and taking as an example the story of flying machines and the life of the Brazilian pioneer Santos-Dumont, Thierry Gaudin (head of the department of technology assessment in the French Ministry of Education and Research) underlines the 'power of the dream'.[11] The life of creators is dedicated to making dreams come true, as Wagner said: 'The great work of the poets is to record and translate their dreams'.

Dreaming

Dreams are, in a way, ancient dreams of humanity and, at the same time, indulgences of the ordinary: everyone dreams. The great reveries of power, goodness and beauty are made of the same stuff as the minor dreams of the loss of money or the chase. Many creators have found inspiration in their dreams: dreaming furnished them, suddenly, critical elements to be integrated in their worries of the moment. *Kubla Khan*, Coleridge's poem, is one of the most famous examples of a creation emanating entirely from a dream. Often, ideas received through dreams provide the raw material of inventions made during wakefulness. It was in this way that Kekule came upon the cyclic molecular structure of benzene.

Creative dreaming is far from being a sporadic, random, uncontrolled (and uncontrollable) activity. Techniques used by the Senoi of Malaysia, North American Indians or Tibetan yogi, aimed at controlling their dreams, correspond to the creative process. [12] Contemporary psychology identifies, in this respect, the following sequential steps: motivation for the creative act, collection of relevant information, then the first attempts at synthesis of these data.

The dreamer learns to relate to his problems while dreaming. Today, Jungian psychoanalysts or 'psychosynthesists' use techniques of mental imagery similar to those of the traditional societies, i.e., directed daydreaming, or dialogue with the superego and the subconscious. Such techniques are pertinent to both psychotherapy and creative effort.

Cultivating awareness

Disciplines of self-knowledge usually focus on awareness and vigilance. 'It is he who, after having been negligent, becomes vigilant, lights up the earth as the moon emerging from the clouds', said Buddha to his disciples. Whatever form meditation takes, its essence lies in awareness and observation. The breathing exercises of yogi aim to develop powers of concentration indispensable to deep understanding, vision and penetration of the very nature of things.

Vigilant awareness of our activities, as preached by Buddha, consists of living in the present, within the act itself, in order to attain Nirvana. The creative illumination finally sparks when the creator, completely absorbed by his creative act, rids himself of self-consciousness.

The civilisations of Asia practice, as those of the West, these disciplines in varying forms. David Bohm, who long worked with the Indian philosopher and spiritual teacher, Krishnamurti, reaffirms the active role that meditation and self-examination play in the abandonment of presuppositions that block our knowledge of reality. 'One can say that Newton, in that sense, meditated because he looked at all different ideas in physics, and these became a part of him'. [13]

Because creators are able not only to observe themselves but also open themselves to the less conscious, many levels of consciousness are made available for their participation within the total process of creation.

Self-organisation

If we reiterate, if we continuously repeat self-observation (in the manner of the bouncing images in carnival games of appropriately spaced moving mirrors), do we risk the loss of self? We are very close to what haunted the classical logicians—who feared the devils of reiteration, repetition and self-reference as sources of paradox and insurmountable contradiction.

The study of autonomy and self-organization, which is currently in the process of developing into a true science,[14] provides an anchor for the notions of autonomy, self-organization and creation based on new conceptual bases. Does not the autonomy of an open system, in fact, appear as paradoxical as the appearance of novelty? Openness to randomness and 'noise', the emergence of novelty and meaning are building blocks (for the French biologist, Henri Atlan) of a system's autonomy. Francisco Varela, the Chilean biologist-philosopher, applies the notion of autopoiesis* extremely successfully to the study of the vertebrate immunological system. He speaks of 'the capacity of an organized system to create and constantly renew its constituents'.

A close relation between processes of autonomy and creativity emerges if we replace *autonomy* by *creativity*, as (for example) in the comments of Jean-Louis Le Moigne, who wrote in the journal *CoEvolution* after the Cerisy congress on self-organisation: 'creation appears as a response to the conflict of chance and necessity . . . creation is the joining of purpose and event . . . there is no creation without memory . . . there is no creation without purpose'.[15]

Complementarities

Narcissus, upon seeing for the first time his image reflected in the water of a fountain, falls ecstatically in love with his reflection. The myth dispels the danger of total self-absorption. The metaphor of the mirror provides, at the same time, the key to escape from the impasse.

Reversing right and left as they do, mirrors suggest complementary roles of the right and left halves of the brain. The left hemisphere corresponds (in right-handed people) to language, verbal activities, reading, writing, rhythms and temporal sequences. To the right hemisphere is attributed recognition of patterns and melodies, spatial activity, non-verbal thought and (for certain researchers and artists) creative action. Pairs of words illustrate these different roles: logical-intuitive, rational-emotional, analytical-synthetic, and so on. For the neurosurgeon, Joe Bogen, the right hemisphere represents 'the world within me', whereas the left activates 'me within the world'.

---

* Eric Jantsch has defined autopoiesis as 'the characteristic of living systems to continuously renew themselves and to regulate this process in such a way that the integrity of their structure is maintained'. (*The Self-Organising Universe*, Oxford, Pergamon, 1980).

The phenomenon of early imitation

This research does not suggest a simplistic dichotomy. Instead of a sharp duality between the two cerebral hemispheres, the findings propose a complementarity of functions that are indissociable. Furthermore, the brain, a living organ, branches out through the whole body. We think and create with the entire body. Neurophysiology teaches us that the structure of the brain is modelled through experience according to the incessant dialogue that one has with his internal and external environments—a dialogue that builds and enriches our cognition. There exists also an internal dialogue that 'drives' mental images to be embedded in memory.

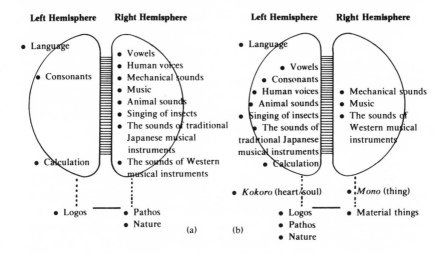

Figure 1. Common conceptions of the loci of sensation or control of various human activities: (a) the functioning of the so-called Western brain, (b) a construction of the functioning of the Japanese brain, as developed by Tadanobu Tsunoda of the Tokyo Medical and Dental University and announced by him during a Unesco-sponsored symposium held in Athens, 1981. © Tadanobu Tsunoda and The Unesco Courier, February 1982, and reproduced with their kind permission.

The creative individual employs the analytical and the fantastic, the mathematical and passionate approaches, without waiting for the neurophysiologists and psychologists to confirm experimentally the details of these complementary functions. Edward de Bono has commented that 'lateral' thinking is useful for engendering ideas and new approaches, whereas logical thinking is needed for their development. Lateral thinking increases the utility of logical thinking, and offers it a great number of choices. Logical thinking multiplies the efficiency of lateral thinking by making good use of the ideas generated. Lateral thinking resembles the reverse gear of an automobile: no one drives continuously in reverse gear, but everyone needs it to maneuver out of impasses.

Complementarity is manifested throughout the creative process: there is first the preparation, then the instant of creation, afterwards the slower process of verification. All creators, even the most inspired, insist on the acquisition of prerequisite knowledge (including detailed information), mastery of the working methods, and—most importantly—continual practice. Inspiration cannot replace work, as Thomas Edison said in somewhat different words. 'The preparation of a work consists of strenuously giving oneself the liberty to perform with gentle deliberation', wrote Valéry. Painters often begin their careers by imitating the works of their masters. Between the ages of ten and fifteen, Johann Sebastian Bach secretly copied the music of contemporary Italian, French and German masters. The same imitative phenomenon has yielded results of undoubted technical inventiveness; the example of Japan is eloquent in this regard.

The role of time

The creative function—with all its unpredictable, irreproducible, unique and irreversible qualities—is deeply embedded in history. The spirit of creation must endure. The moment the light appears is only 'the new embryo', said Valéry, 'alive but not yet living. It must be borne after its moment of conception . . . Lightning itself yields nothing, its flash only illuminates the landscape—beyond—to be explored'.

Memory, practice, continuity and dedication play essential roles in the flourishing of creativity, in every epoch and in all cultures. Central to the current understanding of time are the discoveries of thermodynamics, information theory, cybernetics and biology. These research results weave patterns of time and creativity, echoing the intuition of Bergson's *évolution créatrice.*

The maintenance of the organization of an open system, whether it be a living cell or a human society, requires restraint of the rate of growth of the system's entropy. The creation of a new organization opposes, a fortiori, entropy, stimulating, in Eric Jantsch's words, the 'binding of time'.[16] The French biologist and writer, Joël de Rosnay, has concluded that 'individual creative action compensates for the flow of time; all original works are reservoirs of time'.[17]

The creation of original work thus corresponds to a self-organization of time that influences both the creator and his surroundings. The process then becomes entangled within itself: creativity nourishes the life of the creator, the environment and predicament in which he was born and grew up. His creativity transforms and, at the same time, complexifies these elements of his environment.

A creator creates himself through his works, uniting his internal world with the world without. It is thus, said the great American photographer, Ansel Adams, 'as we grow older, mirrors become windows'. ▪

Notes

1. D. Abram, The Perceptual Implication of Gaïa, *The Ecologist*, Vol. 15, No. 3, 1985.
2. See, e.g. P. Watzlawick, J. Weakland, R. Fisch. *Change, Principles of Problem Formation and Problem Resolution*. New York, Norton, 1974.
3. E. de Bono, *The Uses of Lateral Thinking*. London, Jonathan Cape, 1967.
4. P. Valéry, *Cahiers*. Paris, La Pleiade-Gallimard, 1974.
5. A. Koestler, *The Act of Creation*. New York, Macmillan, 1964.
6. P. Klee, *Théorie de l'art moderne*. Paris, Gonthier, 1968.
7. G. Bateson, *Mind and Nature, A Necessary Unity*. New York, E.P. Dutton, 1979.
8. D. Bohm, Le déploiement de la matière et de la conscience, *CoEvolution*, No. 11, 1983.
9. H. Poincaré, *Science et methode*. Paris, 1908.
10. H. Dreyfus, *What Computers Can't Do: The Limits of Artificial Intelligence*. New York, Harper & Row, 1979.
11. T. Gaudin, *Pouvoir du rêve*. Paris, Centre de Recherche sur la Culture Technique, 1984.
12. See Patricia Garfield, *Creative Dreaming*. New York, Ballantine, 1974.
13. D. Bohm, *op. cit.*
14. See the proceedings of the Cerisy congress, In: P. Dumouchel and J.P. Dupuy (eds.), *L'auto-organisation, de la physique au politique*. Paris, Le Seuil, 1983.
15. J.-L. Lemoigne, Naissance de la science de l'autonomie. *CoEvolution*, No. 6, 1981.
16. E. Jantsch, *The Self-Organizing Universe*. Oxford, Pergamon, 1980.
17. J. de Rosnay, La création collective du temps, *CoEvolution*, No. 7, 1982.

To delve more deeply

SAGAN, Dorion, Sexualité et créativité, *CoEvolution*, No. 12, 1983.

SILVERMAN, B.G., Toward an integrated cognitive model of the inventor/ engineer, *R&D Management*, Vol. 15, No. 2, 1985.

SINGH, Dharamjit, Vision, Its Creativity, Limits and Paradoxes, *Impact of Science on Society*, Vol. 31, No. 2, 1981.

VARGIU, James, Creativity, *Synthesis*, Nos. 3-4, 1977.

VIDAL, Florence, *L'instant créatif*. Paris, Flammarion, 1984.

WOLFF, M.F., To Motivate, Set Goals, *Research Management*, November-December 1985.

A Polish mathematician and control-system theorist induces us to examine the possibilities for a new socio-cultural approach, throughout the cultures and civilisations of the world, to education, ethics, values and respect for others. He uses the Einsteinian device of a 'thought problem,' in which we examine earthly behaviour of ourselves and our congeners through the eyes of others from our own galaxy. The results are cause for innovative thought and action.

Chapter 12

Education for a new cultural era of informed reason

Andrzej P. Wierzbicki

Now a professor at the Institute of Automatic Control in Warsaw, the author recently spent six years working on problems of optimization and decision theory while he chaired the Systems and Decision Sciences Program at the International Institute for Applied Systems Analysis at Laxenburg (Austria). This chapter is based on materials and experience gathered while Dr. Wierzbicki was at IIASA. His address is Institute of Automatic Control, Technical University, Nowowiejska 15/19, 00-665 Warszawa (Poland).

Invitation to a thinking experiment

Most of us would agree that the rate of change in human societies is especially great today and that these changes have a global character. While working at the International Institute for Applied Systems Analysis, I became confronted with diverse perceptions of the concept of global change. A typical emphasis is on the change in economic structures and the global character of trade that intensify international competition and motivate innovation. An even more far-reaching consequence is in the change in technological and scientific potential, or in its impacts on diminishing mortality and changes in the character and speed of demographic development. Other emphases are on the dangers of ecological and environmental change.

All these changes are interlinked, and I shall show that another aspect of global change is still more fundamental: the change of basic cultural values in contemporary societies.[1] These changes will necessitate, in turn, fundamental reform of our educational systems. Before analysing the needs of such reform, we must first understand the main mechanisms of current global change, see them in cultural and historical perspectives, and anticipate the main features of an approaching new cultural era. In doing so, it is extremely difficult to (a) preserve objectivity and (b) not be influenced by the perspective of one's own culture. I ask the reader, therefore, to participate in a mental experiment whereby we put ourselves in the position of social historians.

We are now social historians, but we represent an alien, non-human intelligence. We are studying and trying to understand the last few centuries of development on this planet.

Mechanisms of global change

The alien observer would determine that a sense of joint responsibility for the globe is growing in human societies. This has been caused primarily by the development of mass transport and communication. Even if brought up in one of the nationalist cultures that prevailed in the 19th century—in the chauvinistic belief in the superiority of their own subculture, people have discovered shared values: curiosity, humour, mutual understanding. Human excursions to the cosmic outskirts of the planet and to their moon have taught them that the globe is beautiful and precious, but small and fragile.

The same development of technology that has brought humans close together and improved their perspective has also brought the hazard of global deterioration of environment and climate, especially in terms of potential nuclear war. If nuclear weapons were used by only one 'victorious' side, the 'winners' would only die more slowly and more painfully than the 'losers.' And this is another reason why the human race must learn to live together despite its cultural, ideological and other social diversity; otherwise, the species risks being remembered only as a historical warning to other possible intelligent races inhabiting the Galaxy.

Some draw from this conclusion that technological development is bad. It is even the fashion in many intellectual circles to think so (a phenomenon to be examined more closely later). Yet, this seems unlikely. The species, seen from a historical perspective, does not turn back from dangerous possibilities. The current global socioeconomic system will force upon mankind increased speed in the change of technological and economic structures for another half century or more.

The vast educational transition of the 20th century has brought basic instruction to even the poorest countries. In parallel with this, international capital movements are now much more mobile than before. Seeking cheap labour, capital moves any technology that does not require labour possessing advanced skills from so-called developed countries to industrializing (i.e., poorer) nations. This sharpens the competition for comparable commodities still produced in industrialized countries and creates the need for the application of automation, robotics and advanced information technologies. The demand for labour in the traditional professions and trades declines, but it remains difficult to retrain and re-educate humans quickly enough to compensate.

The resulting 'structural unemployment' and social tensions cause governments to apply measures to slow this process, e.g., restrictions on technology transfer, monetary policies to attract capital back to developed countries. Developing nations are rather unsuccessful in counteracting such policies, and the diversified interests of the international capital market may slow this process even further. Yet the motivating force of such change—the availability of less expensive labour in the developing countries—will remain effective as long as large discrepancies in standards of living persist.

The emergence of over-industrial society

This mechanism of (global) structural change depends little on the differences in national economic systems. Although differing socio-economic arrangements might accentuate one or another aspect of this transition, the so-called planned economies (one type of socioeconomic system) will be forced by international competition to apply similar measures of accelerated technological growth as have done the 'market economies' (a type of system more successful in mass-technology applications). Investments motivated by cheap labour are made from both state-owned and privately held capital. The speed of the transition we are discussing depends, essentially, more on socio-cultural limitations than on socioeconomic factors. And although mankind has witnessed many changes during the 20th century, it is not prepared culturally for such fundamental change.

The scientific-technological foundations of this change include (besides the elements previously mentioned) technical measurement and control, electronics, computers and telecommunications. Science and technology in these fields are prepared to support far-reaching changes, evolution leading to a society that would require relatively little human labour for the production of basic material goods. In the highly developed countries, the agricultural sector now employs less than 10 percent of the labour force; in due time, the respective share of the manufacturing sector will also diminish.

The rest of the labour force will be employed in the development of new technologies and (mostly) in a broadly defined service sector. Depending upon a country's social arrangements, there might also be a considerable number of unemployed. A fashionable theory in some of the countries attributes this trend to an increased demand for leisure. But this is a tautological explanation. The relation of rising unemployment to the mechanism of global structural change, described earlier, provides a much sounder explanation of the phenomenon.

A future human society of the type described is often called, on earth, a 'post-industrial' society. The choice of name illustrates the current cultural limitations of the human race; these impede and distort the changes. The percentage of humans working on new technologies is still rather low, and they tend to concentrate single-mindedly on their work, without reflecting on possible future social developments. Most 'intellectuals' reflecting on the future are not trained in electronics or the use of computers; they do not understand them fully, and they distrust them instinctively.

The coming human society will witness, rather, even more rapid technological development than at present. So we should consider the necessary cultural premises of an 'over-industrial' society.

The end of a cultural era

Few humans realize that they still live under the influence of ideas originating in the 'Age of Enlightenment' (a cultural platform of basic beliefs and values formed about three centuries ago) and that this era might be coming to an end. The technological developments of this period on earth emerged from a specific subculture, the one called European. Enlightenment ideas, such as the concepts of natural law and the natural rights of people, were formulated by members of this subculture tired of religious wars common to the 16th and 17th centuries. European societies began looking forward to new opportunities in the Americas and other regions.

The economic counterpart of these ideas (advanced by one Adam Smith), the belief that the search for individual benefits always benefits human society collectively, was accepted by the people of this subculture under very strict moral limitations (those implied by Puritanism, a variant of Christianity), on the assumption of open frontiers and the possibility of escaping oppression.

Enlightenment ideas motivated, as a consequence, waves of educational reform and new universities in Europe and the Americas; they led to the industrial revolution in Great Britain and other European states and provided a basis for the American and French revolutions. They also led, indirectly, to a new ideological doctrine called Marxism and to the Soviet revolution that implemented this doctrine.

There is an old scholastic dispute among human beings about the origins of chickens and their eggs, as well as about which of the following came first: socioeconomic or socio-cultural development. In the contemporary terms of 'system theory,' humans began to understand that there is feedback between socioeconomic and socio-cultural elements: they influence each other, and there may be periods when one or another is in the ascendant. The cultural elements of the Enlightenment 'platform' contributed to the economic and technical development of the 18th to 20th centuries, an industrial revolution that spread over most of the planet. This socioeconomic development motivates, in turn, ideological confrontations and changes of cultural values.

The challenge of limiting human confrontations

When the early industrial revolution led to new accumulations of wealth and the rise of social inequities, two intellectuals (Karl Marx and Friedrich Engels) protested against these in the Communist *Manifesto*— the basis of Marxist doctrine. After much struggle, this in turn led to the March and October revolutions in 1917 and the ultimate formation on earth of two opposing, major economic systems. Motivated by Marxist doctrine, the centrally planned economies on earth adopted socio-economic arrangements stressing economic equity and collective (as opposed to individual) benefits.

These arrangements have been successful in science and education, but less successful in technological and economic development. The difference in achievements may be attributable to the fact that humans are essentially individualistic. The development of a collective rationality does not come naturally to them (regardless of their socioeconomic arrangements). Yet the confrontation between collectivist and individualist doctrines has been the motive force behind many developments on this planet for the better part of a century.

This confrontation has forced market economies (following essentially the *laissez-faire* doctrine) to abandon classical forms of unrestricted capitalism, to develop new social arrangements that pay off in fast technological development. This led to a competition to conquer space around the planet, and motivated the refinement of electronics and computer technology which, now, are preparing to support a further fundamental change.

Were this confrontation to assume the intensity of the conflicts between Catholics and Protestants (two closely related variants of Christianity) in earlier centuries, it would probably lead to nuclear catastrophe and the end of intelligent life on earth. In the interest of preserving the cultural diversity of the Galaxy's intelligent races, we can only hope that humans will limit the intensity of their otherwise constructive confrontations. But we should watch the humans closely, nonetheless, lest they extend their nuclear confrontations to Galactic space. We grant them the right to destroy their own globe if they so wish; but the pollution of the Galaxy is another matter.

Signs of cultural disorder and decay

While the technological and scientific situation on earth is ripe to support a new era, the Englightenment's cultural platform is no longer sufficient for the purpose; there are many signs of cultural decay. The strict moral limitations guiding individualist rationality at the beginning of this age are now largely dissipated, and permissive societies have developed. [2] Permissiveness concerns not only the mores related to the reproductive behaviour of humans (rather unrestricted and exhibitionist); it is reflected also in the humans' tendency to replace work by entertainment, reducing their sense of obligation towards their fellows, and changed ethics in regard to observing oral contracts. (The need for legal documents implies high contractual costs in any economy.)

Confronted by such dissipation, many humans seek escape in religious activities, and another reason for the turn towards faith is fear of the future—a sense that the old political ideologies provide insufficient guidance. The new possibilities presented at the beginning of this era have been largely exploited, indeed, and the planet now knows a more or less tightly interlinked economic system. So new opportunities lie mainly in the space around the globe, or in exploiting new scientific and technological developments.

The distrust of technology and the 'empirical sciences' has reached the humans' philosophy. After more than a century of stressing the importance of empirical evidence in systems called *positivism, logical empiricism* and *neopositivism*, the earth's philosophers are now turning to the tautology of perceiving science and philosophy as matters of concern to scientists and philosophers. Others,[3] confused by the unresolved dispute between individualist and collectivist beliefs, either (a) condemn any social process guided by vision, and argue that a free society can develop only by vision-free responses to contingencies, or (b) stick stubbornly to a visionary creed, trying to fulfill it at any cost and despite changed conditions.

Both of these fashions in philosophy are further signs of cultural confusion and decay—since an intelligent being knows that true creativity needs vision and an outline of future activity, while implementation of any idea requires corrections and the incitement of further ideas by all individuals participating in the act of creation.

Long cultural waves

Humans have discovered various cyclical, or nearly cyclical, patterns in their economic and demographic development.[4] Such patterns should be seen, of course, against a background of general growth. A hypothesis of a long cultural wave characterizing human development has been put forth, and there are many indications to support it. The cultural era of the later Middle Ages began in Europe around the end of the 10th century with the idea of Pax Dei, i.e., a cessation of warfare among Christians in Europe and the exportation of warfare in its place. This led to the creation of the first universities on the continent (in Italy, France and Great Britain, and in Spain),* as well as to remarkable technical and economic achievements and the prosperity of the late 11th to early 13th centuries.

---

* Earth's Arab civilization, in fact, established the first universities, al-Qarawiyin at Fès (Morocco) in 859 and al-Azhar at Cairo in 970. Only in the 1000s did Bologna follow, and then Paris and Oxford during the 1100s.

Cultural feedback via long 'waves'

'Cultural wave' is used here because we can relate its length to a possible mechanism, namely, changes in basic cultural beliefs. Cultural anthropologists agree that there is a basic delay in changing cultural beliefs, of three to four generations or seventy to one hundred years—related to the fact that, in a family, our beliefs are also influenced by our grandparents. When we ask a mathematical question: what length of wave is possible in a system with accumulation and delay, possible oscillating through a feedback mechanism, we can obtain a definite answer: the typical length is four delay times. Take an equation $\dot{x}(t) = u(t - T_0)$ to account for accumulation and delay T_0 and assume $u(t) = -k.x(t)$ to account for a negative feedback. Consider these equations in frequency domain after the Fourier transformation and apply the Nyquist stability condition. The main transfer function is $G(j\omega) = X(j\omega)/U(j\omega) = e^{-j\omega T_0}/j\omega$, with $\text{Arg } G(j\omega) = -\pi/2 - \omega T_0$; this phase shift must be equal to $-\pi$ in order to generate through the negative feedback a wave of the pulsation

$$\omega_w = \frac{2\pi}{T_w},$$

where T_w is the wave length. Hence $\omega_w T_0 = \pi/2$ and

$$T_w = \frac{2\pi}{\omega_w} = 4T_0.$$

Thus, waves of the length of three to four hundred years, four times seventy-five to one hundred years, are the shortest that can be related to the changes of basic cultural beliefs (at least historically; the pattern might be shortened in the future, because of the impact of global communication and mass media systems).

There are many direct indications of this long cultural wave. Historians of culture agree on three main cultural periods in Europe. One is the culture of the late Middle Ages, starting around the year 1000 with the formulation of the idea of *Pax Dei* in southern France, forbidding warfare between Christians and resulting in the export of warfare through the Crusades, which in turn extended horizons and

prepared for the next period. The second is the culture of the
Renaissance, starting around the end of the 14th century in Italy with
the rediscovery of the banking system along with many basic cultural
achievements of ancient Rome and Greece, and spreading through
Europe in the 15th and 16th centuries. The third is the culture of the
Enlightenment, started by philosophers and scientists towards the end
of the 17th century and spreading through Europe and beyond in the
18th century. All of these changes were followed or accompanied by
major educational reforms.

— Cesare Marchetti

*Reproduced from Evolving Cultural Paradigms, which appeared in
IIASA's quarterly letter* Options (*No. 3, 1984*), *with permission.*

These developments, reflected in part by the Gothic cathedrals
constructed at that time, were motivated by the culture and schools of
the monasteries. As wealth accumulated and social inequalities arose, it
was also the monasteries that initiated intellectual protest. The
Franciscan order, for example, contended that monks be poor and not
rich. Countered and persecuted by an Inquisition instigated by another
order of monks (the Dominicans), the protest movement became radical
and led to much social upheaval.

Stubborn ideological confrontations and other signs of cultural
dysfunction were commonplace,[5] and wars again devastated Europe in
the late 13th and during the 14th centuries. The earlier export of
warfare, however, through a device called the Crusades—large military
expeditions against people of other religions in Asia Minor—broadened
the Europeans' horizons and brought diversified knowledge and new
customs that would prepare the way for the next cultural stage.

The subsequent cultural period in Europe (the Renaissance) is claimed
by the humans' historians to have occurred in the 15th and 16th
centuries. But its economic foundations had been laid in the 14th
century, with the rediscovery of financial credit and the development of
banks in Italy. (The ancient system of credit had been earlier revived by
the Chinese, Indian, Egyptian, Greek, Roman and Arab cultures.) Along
with this came the next wave of European universities, mostly in Middle
Europe: Bohemia, Poland, Germany and Hungary. This prepared the
cultural foundations for a period of remarkable economic and
technological expansion, Gutenberg's new printing technology playing a

key role in the process.* Then came the geographic and scientific discoveries of the 15th and 16th centuries.

The advent of time for change

Three and a half centuries before the Communist *Manifesto*, in the midst of the Renaissance, individual wealth and other new inequalities drove an intellectual named Martin Luther to protest via another manifesto, the ninety-five Wittenberg Theses (1517). This created the variant of Christianity called Protestantism, the genesis of social upheaval and the religious wars that afflicted Europe for another century and a half. During the ensuing Counter-reformation of the 17th century, further cultural disorientation manifested itself when humans went so far as to burn at the stake some of their most original thinkers.

Will humans repeat previous errors and intensify their current ideological stand-off (no more rational than that between Protestants and Catholics three centuries ago), to the point of devastating not only Europe but the planet as well?

According to the theory of the long cultural wave, its duration is specified as three-four centuries and it consists of four delay times[6] during which basic beliefs and values change; the last part of the cultural wave, when cultural confrontations are exacerbated, is about one delay-time long. Before the late 20th century, this delay corresponded to three-four human generations, say 75-100 years. The present finds the earth in the last, dangerous part of a cultural wave.

The development of mass-transport and communication systems on earth might have shortened this span considerably, and this may yet save mankind's civilisation—if humans can change their fundamental ideas and value system in time, re-educating themselves for the approaching new era, and thus avoiding a cataclysmic intensification of ideological disputes. For this purpose, humans will need an accelerated reform of their educational processes, hastening the break-up of their long cultural wave. Other intelligent races in the Galaxy will be watching with excitement.

The age of Informed Reason

Thus ended the report of our alien social historian. When speaking of our proper plans for the future, we need to abandon his objectivity and

---

* The first moveable type, made of wood, was developed by the Koreans and Chinese several centuries earlier.

disdain, for it is our future and that of our children that hang in balance. What lessons should we learn from our own history?

The first imperative is the *unity of mankind's destiny* and the responsibility of the human species therefor. These are needed now in a much deeper sense than in earlier cultural times. The world's destiny means that we must eradicate hunger, control pollution of air and water, and strive to minimize the number of disappearing animals and plants.

A second value involves higher *respect and tolerance for* mankind's *ideological and other cultural diversity*. It is true that during the Second World War the Japanese were often represented as fanatical and cruel warriors. But that nation has many cultural advantages: respect for their elders, love of order, and *sonkei*—a combination of justice, honour, reliability, respectability, and accumulated skills in the service of society.[7] These qualities may have influenced strongly, after the war, the success the Japanese have known in their variant of market economies.

Conflicts have typically escalated[8] beyond rational limits because of incompatibility between the aspirations of the adversaries. Learning about the other side's wishes can mean learning to understand its culture. Culture not respected is culture not understood. Yet tolerance for other cultures need not imply the convergence of cultures or a 'cosmolitan' separation from national traditions. We cannot, at any rate, predict which cultural values mankind might need in the future, whatever their source might be.

There is, next, a third fundamental concept of what is required in a vitally new age. We need a *new perception of rationality*, here called *humane reason*. I see this as a synthesis of the individualist and collectivist perceptions of rationality, going beyond a mere mixture or trade-off between the two. Collectivist and individualist rationalities have been combined since the very origin of cultures. An aboriginal tribe depends on its survival, for example, as much on cooperation and altruism by all its members as on the initiative of its most dynamic individuals. The prolonged era of the Enlightenment was marked by a confrontation between opposing (though simplified) perceptions of rationality. It is only recently that strict theoretical foundations have been advanced for a 'non-trivial' synthesis of these contradictory apprehensions.[9]

The traps of individualistic and collective rationality

Individualistic rationality motivates innovation and growth; it is thus needed in any society. But such rationality has severe limitations, even traps. Two individualists in conflict usually arrive at outcomes that are much worse than if they had been cooperating. Since one never knows when one might be in need of cooperation, the best survival strategy for a human is non-naive altruism. This concept, attributable to Anatol Rapoport,[10] is characterized by willingness to cooperate; this willingness is interrupted temporarily by non-cooperative behaviour 'if the other side cheats,' and then returns to forgiveness and cooperation.

Strong collective rationality has its own traps, one of these being that one can always demand more social justice since there is no absolute 'concept of fairness.' (Absolute fairness would mean that everyone would have the same needs, think similarly, publish the same papers, and so forth.) A society of absolutely equal numbers could not develop further, so that 'humane reason' must reward individual initiative (innovation). And yet such freedom cannot include the liberty to hurt others. Humane reason must, therefore, include non-naive altruism as a behavioural norm. The same altruism implies a readiness for collective ventures, yet limited in such a way that they do not affect basic individual rights.

Fourthly, an important element of the new cultural platform is *respect and care* for our fellow humans. Respect is a demanding relationship: we can demand from our fellow humans that they strive to earn it. Respect may be difficult to develop, but the Japanese concept of *sonkei* shows how effective it can be. Although the notions of humane reason and care for our fellow members of the race may seem very idealistic, it is true that the overindustrial society will require enhanced social ethics.

Fifthly, a vital element of the new cultural platform is *informed humanism*. Conventional humanism is insufficient in today's world because it ignores the verities of technology, including computers, and experimental science; it is uninformed humanism. No one in the overindustrial society will be able to claim to be educated unless he understands what the new technologies can and cannot do. Even the traditional concept of profession will differ from previously (and I daresay that universities may one day cease to grant diplomas).

The final key element comprising the new cultural platform is expressed by the value of *learning, adaptation and creativity*. Structural economic and social changes will force either continuous or discrete changes in the vocations of individuals. They must thus be prepared to learn throughout life ('continuing education' is an idea fostered by Unesco since the 1960s). Especially creative persons may prove to be the only ones capable of remaining in a single occupation for a lifetime.

These six principal elements constitute what I have named the coming era of informed reason.

Education for the coming years

The fundamental changes in education required for the coming era cannot take place rapidly. Although there is no time to waste, we need to envisage two stages—each about a generation in duration.

First stage — Preparation for reform by incorporating the elements of the new cultural platform in existing curricula at school and university levels. Stress should be placed on:

- Evolving cultural features of the future, especially improved ethics and the adoption of higher degrees of responsibility than heretofore;
- Informed humanism, with wise instruction in the foundations and uses of data processing, system theory and automation.

Second stage — Concrete reforms in schools and universities. This implies realization, in practical teaching terms, of the philosophies cited above. The aims are to improve the quality of, and access to, education. Some economists believe that the industrialized countries have already reached the point at which they are overeducating their citizens.[11] But this short-term view of the 'market value of human capital' does not explain the phenomenon of human curiosity (the propensity to learn much more than is needed for immediate application). This phenomenon indicates that education should be understood rather as a value for the future of the human race than as an individual market value.

The quantity-quality dilemma

Increase in the citizen's access to education tends to put in question a concomitant rise in the quality of education available. A simplistic resolution of this dilemma, or paradox, is to accept a rise in quantitative access with an acceptance of a reduction in quality. My own experience, while participating in or initiating, educational experiments designed to solve this problem is as follows.

- Open access to university should not be understood as the right to enter without qualifications or competition. Objective, demanding entrance examinations are necessary to maintain quality while increasing quantity. Open access can be realized through extension training: evening classes, correspondence courses, other continuing education.

- The aim of raising the chances of socially handicapped persons to enter university can be done through a two-stage system of admission, a system that I originated while serving as dean of the Faculty of Electronics at the Warsaw Technical University: a small number of candidates passing the entrance examination are admitted immediately to normal studies, while candidates with lower marks are admitted conditionally. The latter must begin working in a field associated with the study area while following evening courses covering the first year's—really, first semester's—curriculum. Those who pass the demanding examinations then move to the next year (or semester) for further normal studies. Twelve years' experience has shown that the procedure (a) adequately tests motivation to work hard and (b) has raised the admissibility of the socially handicapped—children of labourers and farmers or youths from provincial schools.

We need to wait a generation before implementing the next stage of educational reform. During the intervening period, it is mandatory to review the disciplines under study, prepare new university departments, rethink the entire role of university teaching, undertake new pedagogical experiments (see box), extend considerably the system of continuing instruction for both teachers and the socially handicapped, and to begin all these things now. We can make, as we proceed, the required adjustments to cultural beliefs and values.

Vision, I repeat, is essential to a sound and vital culture. But vision can only influence the way that people perceive problems; it cannot solve problems. Problems are solved, rather, by the motivation, innovativeness and will of people concerned with an issue as complex as the improvement of education. This will, in turn, contribute to the chances of survival of our species.

Notes

1. See A. Wierzbicki, Changing Cultural Paradigms, *Options* (IIASA), No. 4, 1983, p. 10.

2. See F. Hirsch, *Social Limits to Growth*. Cambridge, MA, Harvard University Press, 1976.
3. See G.R. Little, Social Models: Blueprints or Processes? In: J. Richardson (ed.), *Models of Reality: Shaping Thought and Action*. Mt. Airy, MD, Lomond Publications, 1984, p. 101.
4. See, e.g., R.G. Wilkinson, *Poverty and Progress: An Ecological Model of Economic Development*. London, Methuen, 1973.
5. A competent description of these times can be found in a philosophical book disguised as fiction, U. Eco, *The Name of the Rose*. New York, Harcourt Brace Jovanovitch, 1984.
6. 'Delay time' is a concept in system dynamics that denotes the time elapsing between a causal action and the first observed effects of this action on the system's output. In the case of basic cultural beliefs, this delay can be measured as the time during which an immigrant family maintains the cultural values of its parent country, or the time required for an idea proposed by intellectuals to penetrate (through family and school) into basic societal values. Until the late 20th century, this delay lasted three to four human generations.
7. Mr. Tetsuzo Tanino, private communication.
8. A. Wierzbicki, Negotiation and Mediation in Conflicts, I: The Role of Mathematical Approaches and Methods, In: H. Chestnut, *et al.* (eds.), *Supplemental Ways of Improving International Stability*, Oxford, Pergamon, 1984.
9. See A. Rapoport, The Uses of Experimental Games, In: M. Grauer, M. Thompson, A. Wierzbicki (eds.), *Plural Rationality and Interactive Decision Processes*. Berlin, Springer Verlag, 1985.
10. *Ibid.*
11. R.B. Freeman, *The Over-Educated American*. New York, Academic Press, 1976.

PART III

Innovation in the
Austro-Hungarian Empire

EDITORIAL NOTE

Innovation in the Austro-Hungarian Empire
Cultural casebook

Rilke? He was a German poet or, no . . . he came from Prague. But wasn't that Kafka? The actor, Eric von Stroheim, was a minor German aristocrat who liked to portray characters from his country's nobility. Victor Adler, a luminary of the German socialist movement during the 19th century, and his son Friedrich—a physicist who dabbled in politics and got involved in an assassination: they were either German or Swiss. Baroness Bertha von Suttner was a Swede who won an early Nobel Prize for peace, and Lise Meitner was a German scientist of some kind. Doppler, Pauli and Schrödinger were distinguished German men of science of the 19th and early 20th centuries. Marcel Breuer: was he not a French architect (probably from Alsace, given the German family name)? And the well-known creator of the literature of management, Peter Drucker, is an American.

Rainer Maria Rilke, Erich Hans Stroheim, Franz Kafka, the Adlers, Countess Bertha Sophie Felicita Kinsky von Chinic und Tettau who married a Baron von Suttner, Lise Meitner, Christian Doppler, Wolfgang Pauli, Erwin Schrodinger, Marcel Lajos Breuer and Peter Drucker are all sons and daughters of the Austro-Hungarian Empire of the Habsburgs. In the chapters that follow, we open wide windows on the creativity and invention that marked the acme of a dynasty extending almost six and a half centuries, from 1278 until its almost total social entropy in 1918.

A virtually incredible synergism

The Austro-Hungarian domains included, besides Austrians-Tyroleans-Hungarians: Bohemians, Moravians and Slovaks (today's Czechoslovaks); Silesians and Ruthenians (of contemporary Poland and the Soviet Union); Slovenians and other nationalities constituting today's Yugoslavia; and a variety of even more ethnic groups less well known on the world scene. All of these cultural entities acted synergistically to advance the cause of civilization—occasionally committing serious societal blunders along the way—with the cosmopolitan capital of Vienna as their brilliant focus.

One of the smallest countries has a remarkable record of inventiveness and accomplishment. Here are traced nearly two centuries of varied and unusually progressive innovation in all sectors of human activity, from the end of the reign of Maria-Theresa until the occupation of Austria by the Third German Reich. Some possible reasons for this amazing level of creativity are advanced.

Chapter 13

Multi-cultural influences on the innovation process: Austria-Hungary and the industrial revolution

Jacques G. Richardson

This chapter has been revised, adapted and combined from J. Richardson, 'L'influence culturelle, politique et économique sur le processus de l'innovation: le cas de l'Autriche, la révolution industrielle et ses conséquences' and 'Origines probables du phénomène de l'innovation', in Innovation et société: l'engrenage des innovations *(Proceedings of the European Academy of Arts, Sciences and Humanities, Paris, 1981), with permission. When the name of an Austro-Hungarian subject, native or naturalized, appears for the first time in the text, it is italicized.*

New departures in innovation

Roughly at the same time that James Watt patented his variant of the steam engine, the French military engineer, General J.-B. Vaquette de Gribeauval, was analyzing for the Empress *Maria-Theresa* problems related to the modernization of the Austro-Hungarian armies. *Josef Haydn*, the son of a wheelwright in Rohrau, was beginning to transform the newly codified form of the sonata into what we now call the classical symphony; his dynamic and long movements, replete with expressive, personal interpretations, used different keys and themes to build tension and contrast. Haydn, sheltered from the world as orchestral director in the employ of the *Esterházy* nobility, probably realized that he was estranged and thought himself compelled to be original. Using new order and form combined with delicate craftsmanship, Haydn helped pull Western music from its less 'absolute' forms of the late baroque exemplified by Bach, Händel and Gluck, to concert-designed compositions of enduring value.

Innovation in Austria and Hungary took many forms from the dawn of the Industrial Revolution onwards. Within a decade of the pioneering by Haydn of the expansive symphony, Maria-Theresa's successor, her son *Joseph II*, undertook numerous reforms intended to provide public contentment as well as a healthy state. Serfdom was abolished; Vienna, the capital, represented the Bohemian, Hungarian, Slovak, Galician and other ethnic strains of the empire; German was decreed the uniformizing, official language; and major reforms took place in administrative procedure and the civil service. Torture was abolished; Protestantism was tolerated and its religious services recognized; and the Jewish population was finally emancipated. These politically novel changes, instigated by a benevolent and reforming sovereign, led nonetheless to sharp Hungarian and Belgian protests which would do irrevocable harm to the Habsburg suzerainty. One man's centralization became others' cause for centrifugation of the monarchy.

Joseph II and *Mozart* died a year apart, by which time the latter's musical enterprise had discarded baroque modes and modernized the Italian opera, first consolidated around 1600 by Monteverdi. (The latter's *Orfeo* was first performed outside Italy in Austria, at Salzburg in 1614.) After *The Marriage of Figaro* and *Don Giovanni*, first performed in Prague, *The Magic Flute* was the earliest significant opera sung in German and derived from Josefinian 'enlightenment' (*Aufklärung*) and the contemporaneous Viennese theatre—including the Singspiel dating from 1760.*

While Haydn had succeeded in responding to the musical demands of the eighteenth century's *ancien régime*, Mozart as innovator finally perished from—among other contributing causes—the incongruity between the cultural attitude of his contemporaries (still largely unreceptive to novelty) and the composer's joyful and seemingly uncomplicated personal equilibrium. But Mozart helped set the scene for the era of the liberated, independent creator, the free artist typified by Beethoven. Although a German, partly of Flemish origin, Beethoven was strongly influenced by Mozart and the eminent Haydn—the three knew each other—and Beethoven's main creative period, from 1793 until his death in 1827, is intimately associated with the rich and enduring cultural life of Vienna.

Beethoven, who developed symphonic thought to its highest musical and intellectual level, set the mood and evolution of the work to follow as well as the symphonic pattern that would be used until the time of *Anton Bruckner* and *Gustav Mahler* later in the nineteenth century. Beethoven's fame pushed *Franz Schubert*, who perfected the *Lied*, a melodic unity of voice with piano, to elaborate this fundamentally lyrical idea, developing further Beethoven's concept of contrasting themes. The political atmosphere during this period was the repressive regime of *Franz I*, a nephew of Marie-Antoinette, who was both the last emperor of the Holy Roman Empire and the first emperor of Austria. His imperial reign was beset by difficulties of political and military nature; it remained impervious to liberal reform, and nurtured the personal hostility of *Klemens* Prince *von Metternich-Winneburg* to the notions of freedom and self-determination.[1]

Despite this oppressiveness in the political sphere, there flowered in Austria first the Empire and then the *Biedermeier* periods of design in the crafts and applied arts. Flourishes from the baroque and rococo periods diminished and disappeared, a highly romantic genre gave way to simpler lines in furniture, porcelain and glass, and occasional Asian influences and even a few from classical antiquity affected

---

*The relative quality of originality is detectable even in Mozart's work. The *Concerto in C-major* of Antonio Salieri, especially its piano sequences, sounds like but is not Mozart. The *Don Giovanni Tenorio* of Giuseppe Gazzaniga (1743-1818) appeared in 1786, a year before Mozart's and DaPonte's own *Don Giovanni*.

Fig. 1. Austria-Hungary of the Habsburg era, until 1918.

ornamentation in silver and china, carpeting and fabrics. The music of court and concert-hall moved into the home, and the new demand for *Hausmusik* helped make the name *Bösendorfer* one of the world's foremost pianos. Well-crafted jewelry was worn, not ostentatiously as at court, but by theatre-goers attending a satirical comedy by *Johann Nestroy*, a grand drama by the tragic poet, *Franz Grillparzer*, or the fairy-tale melancholia of *Ferdinand Raimund.*

Developments elsewhere in the culture

Let us leave the non-scientific world for a while, then return later to the arts, crafts and humanities. Despite a small population (Austria today numbers less than 8 million inhabitants), the country's contributions to knowledge have been significant to the point of seeming disproportionate; they reflect also the past glories and much larger population of the Habsburg empire which attained more than 50 million (Fig. 2). Here we endeavour to summarize the main achievements of this remarkable Middle European nation, standing not only at the crossroads of contemporary East and West but at the junction of the Mediterranean and the Nordic cultures as well as at a nodal point between the

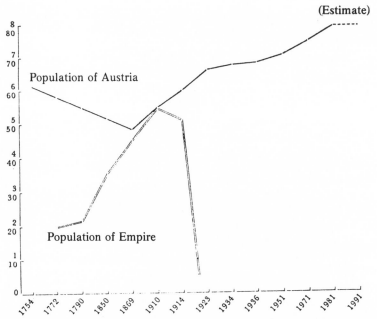

Fig. 2. Austrian demographic growth during and after the Industrial Revolution (From: Institut Autrichien, Paris).

civilisations and religions of western Asia, Islam, the Hebraic tradition and Christianity.

The companion disciplines of physics and astronomy are rich with Austrian accomplishment. *Josef Johann* and *Karl von Littrow*, father and son, founded the Viennese school of astronomy. Josef Johann did much to popularize the celestial sciences; his book, *Die Wunder des Himmels*, first published in 1834, enjoyed publication of a tenth edition as late as 1939. The name of *Christian Doppler* is well-known to students of acoustics and optics. *Josef Loschmidt*, whose constant made possible the first calculations of the weight of the molecule and atom, studied the diffusion of gases. His contemporary, *Josef Stefan*, (Fig. 3), who founded the Austrian school of physics, determined the conductivity of heat in gases, formulated the law of the intensity of thermal radiation for black bodies, and championed atomism. *Ernst Mach*, a philosopher and psychologist, left us the unit of measure for the speed of sound in an aerodynamic medium; his critical views on Newtonian mechanics had considerable influence on Einstein's thinking; and the subjective idealism of his 'empirico-critical' views, contesting the opposition between psychical and physical circumstances, drew strong written criticism from Lenin in 1909.

Fig. 3. An Austrian postage stamp issued in 1985, a philatelic commemoration of the great physicist, Josef Stefan (1835-1893).

Ludwig Boltzmann calculated molecular speeds in 1868 and linked, in 1877, entropy with probability; one can state that it was he who refined the explanation relating thermodynamics to mechanics. One of Boltzmann's students, *Friedrich Hasenöhrl*, killed in the First World War, was inadvertently a precursor of Einstein by way of his ideas on the radiation of moving bodies. A student of Hasenöhrl, *Erwin Schrödinger*, was to win the Nobel Prize in 1933 for his application of wave mechanics to the atom. *Lise Meitner*, the only woman in this pantheon of physics, isolated proactinium in 1918, together with the German Otto Hahn; with her nephew, *Otto R. Frisch*, she interpreted the nuclear fission in uranium by using a nuclear model proposed by Niels Bohr. During the same period, *Paul Ehrenfest* and *Hans Thirring* were major figures in Austrian physics; Thirring, who helped found the European Laboratory for Particle Physics (CERN), also became internationally active as a pacifist until his death in 1966.

But the Austrians were not the only ethnic group in the old empire who contributed to the culture. The Hungarians (whose young noblemen wrote the first non-classical Hungarian poetry while attending the Collegium Theresianum, an elite secondary school in Vienna), made noteworthy gifts to the natural sciences. In physics, *Theodor von Kármán* pioneered in the mechanics of turbulence and in flow dynamics, elasticity and resistance in materials, besides attacking theoretical problems in hydrodynamics, aerodynamics and thermodynamics. He co-founded what is now the Jet Propulsion Laboratory at the California Institute of Technology. *Léo Szilard*, upon leaving Europe to establish himself in the United States after Hitler and the Nazis had come to power, worked closely with Enrico Fermi to develop the Manhattan Project's nuclear reactor program. *Eugene P. Wigner* is a chemical engineer who made major contributions to physics, having shared the Nobel Prize in this discipline (1963) for his discovery and application of the principles of fundamental symmetry.

Victor Franz Hess made balloon trials to establish with Gockel of Switzerland the existence of cosmic rays, universally recognized by 1926, a discovery leading to the Nobel Prize in 1936. *Wolfgang Pauli* helped to develop quantum field theory, elaborated the Exclusion Principle (in 1925), hypothesized with Fermi the existence of the neutrino (1931) and earned the Nobel Prize in 1945. *Hans Edward Suess*, a Viennese physicist and chemist, assisted in the development of the theory of the structure of a layered nucleus (in 1950), thus explaining the existence of the so-called magic numbers.

Chemistry and medicine

The Austrians were not left behind in the chemical sciences. *Friedrich Emich* blazed the trail in inorganic micro-analysis and took the Lieben Prize in 1911, while the work of *Fritz Pregl* in developing the micro-analysis of organic substances won him the Nobel Prize in 1923; and *Richard Zsigmondy* worked on the chemistry of colloidal surfaces, co-developed in 1903 the first ultra-microscope, and in 1925 won the Nobel Prize. Three other outstanding chemists of the past century were *Otto Krathy, Hermann Mark* and *Fritz Paucth*.

In medicine, *Josef Hyrtl*, the son of an oboeist in the employ of Prince Esterházy, modernized the study of anatomy. In 1847, *Ignac Fülöp Semmelweis* established the communicable nature of puerperal disease, or uterine infection, but his analysis of the cause of death in distressful agony by many women after childbirth was ridiculed by incredulous colleagues and the dogma of contemporaneous practice.[2] Semmelweis died insane and a martyr to clinical research. *Sigmund Freud*, physician and neurologist, made behavioural analysis respectable, but only after surmounting barriers almost as high as those confronted by Semmelweis—and in the process became the father of psychoanalysis.

Julius Wagner von Jauregg combatted paralysis in patients suffering from syphilis, developed an inoculation against malaria, and won the Nobel Prize in 1927. *Karl Landsteiner*, a serologist, established blood groups in 1900 and the rhesus factor (in 1940, in the United States), and won the Nobel Prize in 1930. For his work on the vestibular apparatus and spasms of the eye muscles, *Robert Bárány* was awarded the Nobel Prize in 1914. Another *Frisch, Karl von,* shared the Nobel Prize in 1973 with Nikolaas Tinbergen and still another Austrian, *Konrad Lorenz*, for their investigation of zoosemiotics, or communication codes used by animals, and the general field of ethology (animal behaviour).

Among the other life sciences, *Caspar Graf von Sternberg* heads a line of important researchers because of his work on plant fossils, botany and palaeoethnology—a position comparable to that of *Franz Unger* in the reform of the botanical subdisciplines. A widow who began her travels at the age of 45, *Ida Pfeiffer* toured the world in pursuit of zoological and ethnological data which she donated to the Naturalienkabinett and the British Museum. Other biologists include *Johann Gregor Mendel* (see box), an Augustinian monk and the founder of genetics and modern biology, who enunciated the laws of hybridation and genetics in 1866; *Edward Suess*, a geologist and man of politics, who explained marine unconformity by eustacy—variations in sea-level resulting from glaciation and thaw; the botanist *Anton Kerner von Marilaun*, author of the major work, *Pflanzenleben* (Plant Life); *Julius Ritter von Wiesner*, a leading plant anatomist and physiologist; *Hans Molisch*, a master of microscopic and experimental botany; and the father-and-son combination of *Richard* and *Fritz von Wettstein*, specialists in phylogeny.

In diverse biological subdisciplines, *Othenio Abel* was a student of landscape architecture, geology and mining who helped found palaeobiology. (Later he became an extremely active Nazi.) *Albert Szent-Györgyi*, a biochemist from Hungary who discovered vitamin C in the late 1920s, helped isolate vitamin B6 in 1934 and investigated vitamin B3, work which earned him the Nobel Prize in 1937. The physiologist, *Hans Selye*, who later became a Canadian citizen, was in the vanguard of research on stress; he coined the expression 'eustress,' signifying the positive pressures in our daily lives.

Earth sciences and mathematics

Turning to the disciplines related to the earth's crust, the innovative tradition of the first cartographer of the Amazon River (an Austrian Jesuit, *Samuel Fritz*) was to be maintained in geology and other earth sciences by nineteenth- and twentieth-century Austrian specialists. Scientific mineralogy was founded by *Friedrich Mohs*, who established the ten-degree scale of hardness (*Mohs'sche Härteskala*) for classifying minerals by their external characteristics. *Wilhelm von Haidinger* was the founder and first director of the Imperial Geological Institute; he discovered dichroism (the property of seeming color variation in certain substances). *Gustav Tschermak von Seysenegg* created an institute and a periodical devoted to research in mineralogic petrology, and his investigation of complex stone-forming silicates became exemplary in the field. *Friedrich Becke* endowed with his name the optical methods for exact determination of stone-forming minerals, especially those in the feldspar group. And the above-mentioned Edward Suess was the author of the first general geology of the globe, 1893-1901.

In the area of Arctic exploration, *Carl Weyprecht* and *Julius von Payer* distinguished themselves in an era when polar research could have held little interest for almost landlocked Austria. An early specialist in synoptic meteorology was *Julius von Hamm*, who analyzed the *Föhn*, the south-north wind originating in the Sahara, and explained its thermodynamic characteristics. His work led later to better under-standing of the formation of clouds and precipitation.

In mathematics, the Hungarian *János Bolyai* developed a non-Euclidean geometry, contemporaneously with Gauss in Germany and Lobatchevski in Russia, work which was accomplished in the year 1831. Perhaps the greatest mathematician produced by Austria was *Wilhelm Wirtinger*, who did early work on unsolved aspects of Riemann theory (concerned with problems of surfaces), resolving these during the 1890s while teaching at the Technische Hochschule. Exactly a century after Bolyai, *Kurt Gödel* enunciated a theorem whereby it is impossible to demonstrate a hypothetical deductive system on the basis of the system itself. By mid-20th century *Ludwig von Bertalanffy*, a biologist and latter-day Leibnizian, was one of the originators of general system theory—combining integrative philosophy with the flexibility of the Austrian impressionism derived from Vienna's turn-of-the-century psychologists.

For hundreds of years, since the time of Charles V and the rival realm of Philip II of Spain, the forested Austria of the Habsburg dominions served as a political node rather than an economic junction in Central Europe. (In time, Silesia and Bohemia would be the first regions of the Austro-Hungarian Dual Monarchy to industrialize.) There was a mean density of about twenty people per square kilometre; today this average is ninety. The total population of Austria proper in the two centuries between 1754 and 1951 hovered at about 6 million (Fig. 2). Population growth and urban development centered on the capital city, though the fastest-growing city in nineteenth-century Europe was Budapest.

Fiscal and other factors

One of the reasons why the industrial revolution crept eastward through Europe as slowly as it did was the tax barriers common to the Austro-Hungarian, Holy Roman and Ottoman empires. In Central Europe, many of these obstacles disappeared during the fifty-five years separating the Congress of Vienna and the Franco-Prussian War. One result of the political uprisings in Austria-Hungary in 1848 was the removal two years later of the customs frontier which had set Hungary fiscally apart from Austria.

This was a significant economic event, followed in 1867 by a political move of great purport for the future of the innovative process in the country: the Hungarians won equal status with the Empire's German-speaking peoples by virtue of the *Ausgleich* (Compromise). This was the same year in which Bismarck formed the North German Federation after the Prussians had defeated the Austrians at Sadowa (Sadová, in today's Czechoslovakia), and the Habsburgs forfeited their claims to the leadership of the German-speaking peoples. Prussia thus also confirmed its economic ascendancy over Austria-Hungary. Twelve years later, the Dual Alliance tied Austria to Germany—a tenuous bond which was to endure, with time out for the Austrian Republic of 1918-38, until 1945. The economies and cultural outlooks of the German and Austrian commonwealths became intermingled, if not outrightly congruent.

In the east, the Russian search for expansion brought the influence of the Tsar into the Ottoman Empire, imperiling the Austro-Hungarian economic outlets in the Balkans. As the sixth century of Habsburg rule drew to its close, the Austrians also foresaw encroachment on irredentism (as in Serbia and Montenegro), enjoying the protection of the Russians. Austria then sought its safety in attempting to gain political and economic control in the Balkans, as well as creating power blocs to counter Russian imperialism.

Community planning and other applied arts

Just as Haydn, Mozart, Beethoven and Schubert had led Viennese music away from the baroque, so would a new architecture and town planning permit the Austrian capital to leave behind eighteenth-century design concepts elaborated by *Lukas von Hildebrandt, Johann Bernhard Fischer von Ehrlach*, his son *Josef Emanuel* and *Nikolaus Pacassi*: their palaces, churches, libraries and romantic vistas. This was followed in the nineteenth century by the middle-class Biedermeier period cited above, with its Lanner waltzes, 'romantic idealism' and the coming of the music by the Strausses. The Biedermeier, whose end was marked by the savage sacking of Vienna by sons of Austria-Hungary the night of 31 October 1848,[3] served to introduce more than four decades of Ringstrasse Vienna.

Austria of the Ringstrasse period has tended to have historical significance as one of frivolous monument-building (fed at first by the financial boom of the *Gründerzeit*, an age of developmental speculators), the tragedy of Mayerling in 1889, the subtlety of the poet and librettist *Hugo von Hofmannsthal*, the tantalizing dramatic devices of *Arthur*

Schnitzler (sometimes called the Freud of the theatre), the crisis-of-identity literature of *Franz Kafka*, the courage of *Gustav Klimt, Egon Schiele* and *Oskar Kokoschka* in their undertakings of novel genres of painting. But these led to major new cultural vogues: Jung Wien, the café literati; the Wiener Secession, a break-away group from the imperial art academy, and its analogue in the graphic arts, the Wiener Werkstatte (Figs. 4 and 5); and the architecture of *Otto Wagner, Josef Olbrich* and *Adolf Loos*. Thus were born expressionism, the Thonet furniture of bent wood, the musical atonality of *Schönberg, Webern, Berg, Wellesz* and their 'second Vienna school.'

Fig. 4. Couple and Cat in the Snow (colour lithograph) by Fritzi Low-Lazar, an illustration executed in the style typical of the Wiener Werkstätte, c. 1914.

There were, in parallel, major sociological influences of the time, personified by: *Georg von Schönerer*, an opponent of progressivism, a pan-German and antisemite; *Karl Lueger*, a good administrator and friendly opportunist who accepted the mayoralty of Vienna while integrating a limited anti-semitism with the country's political brand of Roman Catholicism; and the Budapest-born journalist *Theodor Herzl*,

*Fig. 5. Bringing Home the Christmas Tree (colour lithograph) by Carl Krenek;
Wiener Werkstätte, c. 1910.*

whose militant utopianism gave birth at first to Zionism and ultimately
to the Jewish state of Israel. Two other utopians were *Theodor Hertzka*,
an economic journalist who published the futurist novel *Freiland: Ein
soziales Zukunftsbild* (Freeland: A Social Image of the Future), and
Martin Buber, who spent a lifetime struggling against racism. Good, bad
or otherwise, all these views and moods augured imminent political
change.

Appreciation of Ringstrasse Vienna needs to take into account the
enormous technical innovation which characterized the nineteenth
century—before, during and after the incredibly long reign of the
Emperor *Franz Josef*, from 1848 to 1916. Franz Josef was a monarch who
had little interest in science and technology but, during his sixty-eight-
year rule, invention and engineering of a scope surpassing those of other
small nations would project their influence well into the present century.
Only political events, it seems, were to thwart the process.

The new reach of technology

About forty years before an American named Singer could
commercialize a sewing machine, a tailor from Kufstein in the Tyrol,
Josef Madersperger, invented the first device of the kind; but the
innovator had insufficient funds to patent or market the machine. A
Bohemian forester named *Josef Ressel* invented the ship's screw, and
Simon Ploessl developed a microscope in 1830 which became the
foundation of the Austrian optical industry. *Ignaz Bösendorfer* and his
son *Ludwig* developed the famous line of pianos bearing the family
name, that which adorned the instrument preffered by *Franz Liszt. Alois
von Negrelli* came from the portion of Tyrol now part of Italy; both civil
and hydraulic engineer, he built the first mountain railways in Austria
and Switzerland and conceived the basic plan for the Suez Canal,
accepted in 1856 by Ferdinand de Lesseps in Paris. *Carl von Ghega* was
another railway builder who conquered the Semmering Pass to Italy
and designed the Baltimore and Ohio railroad in the United States.

From Graz in southeastern Austria came *Anton Schrötter von
Kristelli*, a chemist and physicist who developed a non-toxic form of
phosphorus for use in the match industry. Kristelli was the Emperor's
private tutor in the natural sciences and helped prepare the
Austro-Hungarian north-polar expedition which, in turn, honoured the
scientist by naming Cape Schrötter on Franz Josef Land after him. A
tutor of the young Crown Prince Rudolf was *Siegfried Marcus*, also an
inventor in the fields of telegraphy and automotive engines.

The Uchatius process for alloying steel and bronze in field guns was
named after *Franz Freiherr von Uchatius*. The Glockenspiel in Munich
was built by *Christian Reithmann*, who may have invented the first
four-stroke internal-combustion engine during the early 1860s—at about
the same time that *Wilhelm Freiherr von Engerth* developed
multi-wheeled locomotives to increase the friction surfaces necessary for
good traction on alpine ascents. *Johann Kravogl*, a locksmith from Lana,
devised a system for propelling prime movers by hot air, later adapted to
Paris street locomotives. A major tunnelling engineer of the time was
Franz von Rziha, who replaced the timberwork in tunnel construction by
steel-arch framework. This defense against the pressure of rock became
universally employed.

In hydrodynamics, the naval architect *Fritz Franz Maier* gave his
name to the Maier-Form in ship-hull design, whereas the Hungarian
David Schwartz or *Dávid Schwarz*, through the self-study of aeronautics,
developed the idea of an airship covered by aluminum sheets—a patent
purchased by Graf Zeppelin and integrated into dirigible design. The

engineer who supervised construction of the scenic cable railway up the Leopoldsberg outside Vienna, *Gustav Lindenthal*, later built a major bridge at Pittsburgh in the United States, planned the construction of the Brooklyn Bridge, was consultant on the building of railway tunnels under the Hudson and East Rivers at New York, and in 1917 erected the Hell Gate Bridge in the same city. Another enormously capable civil engineer was *Josef Melan*, who incidentally married the daughter of the above-mentioned Rziha. In his four-volume book, *Der Brückenbau*, as well as in his *Handbuch der Ingenieurswissenschaften* of 1896, Melan insisted upon analysis of the deformation of materials used in bridge-building and on the adaptation of ferroconcrete construction for bridges. This resulted in the successful erection of high-arch bridges at Ljubljana, Genoa, New York, Minneapolis and San Francisco. And, in this vocation, the mixture of Portland cement and asbestos called 'Eternit' was invented by a brewer from Oelmutz named *Ludwig Hatschek*.

In photography, a photochemist by the name of *Josef Maria Eder* exploited his own field as well as those of spectrography and sensitometry in order to advance reproduction techniques, while the chemist, Baron *Carl Auer von Welsbach*, developed the incandescent gas lamp (in 1885), the osmium lamp and the ferrocerium alloy used as the 'flint' in pocket cigarette lighters. In steamfitting, *Hans Hoerbiger*, the father of the famous actors Attila and Paul Hoerbiger, replaced leather with steel in the Hoerbiger valve. And a year before the brothers Auguste and Louis Lumière successfully demonstrated their cinematograph in Paris, the Viennese painter *Theodor Reich* projected 'moving' photographs on film which unfortunately burned during the performance.

New science transforms technology

One of the parents of aerial photogrammetry was *Theodor Scheimpflug*, an officer in the Austro-Hungarian merchant marine and in the army. He developed both instruments and techniques for making photogrammes and evaluating them topographically with the aid of light. Scheimpflug's work led Austria to be the first country to found a photogrammetric society (1907) and, at the same time, this inventor published a major guide to producing cartography via photography, *Die Herstellung von Karten und Plänen*.

A co-developer of stainless steel was *Max Mauermann*, while *Viktor Kaplan* has left his name in turbine technology; Kaplan was able to devise a hydraulic turbine adapted especially to low and variable 'heads' of water pressure. *Ernst Leo Schneider*, who was born in Moravia, contributed his talent and name to the Voith-Schneider multi-directional ship's screw, soon used worldwide on flat-hull vessels. In

aeronautical engineering, *Igo Etrich*, another Bohemian, studied the
wings of birds, bats and insects in order to build his first glider in 1903;
later, he founded the Etrich-Flieger-Werke in Libau, Silesia, where his
chief technical advisor was Ernst Heinkel (later a principal aircraft
designer of the Third Reich).

To end this section on technical innovation and bring us to the modern
era, it is to be noted that the mundane postal cheque was devised in
Vienna in 1882 by a German (Georg Coch) and soon adopted throughout
Europe. In 1931, *Friedrich Schmiedl* launched the first mail rocket,
carrying 102 items of post from the top of Mount Schöckl near Graz to
the valley below. *Gustav Tauschek* developed an automatic bookkeeping
machine and raised the efficiency of moving tools; his bookkeeping
instrument was the first device that could read, write and calculate;
Tauschek succeeded in selling 168 of his patents to International
Business Machines (IBM). *Thomas Gold* and *Hermann Bondi* both still
alive in the late 1980s—postulated with Briton Fred Hoyle the
ill-favoured 'steady state' theory of the universe. But astrophysicist Gold
also predicted to NASA that the moon's surface is covered with fine dust
instead of hard rock. His latest theory is that most of the earth's
hydrocarbons are derived from non-biological substances, a hypothesis
that could have economic significance of global and long-term
dimensions.

Evolution of the social sciences

This rapid scrutiny of cultural evolution, the history of science,
industrialization and economic growth, and the progress of mankind
within a small nation-state culminates with some of the advances made
by Austria in the economic and other social sciences. H.H. Gossen, an
early nineteenth-century German, strongly influenced the ideas
concerning value and marginal utility espoused by the Austrian *Carl
Menger*, the Englishman W.S. Jevons, the Frenchman Léon Walras and
the Italian Vilfredo Pareto. Two Austrian disciples of this Vienna school
of marginal utility (*Grenznutztheorie* or *Marginalismus*), *Eugen Böhm
von Bawerk* and *Friedrich von Wieser*, were to become reformist
economic ministers of their country by the time of the First World War.

An important disciple of these two economists would be *Joseph Alois
Schumpeter* who, in tracing the cyclical development of capitalism,
insisted on the decisive role of the entrepreneur and the major
significance of innovation in economic growth.[4] (See box.) Schumpeter
foresaw the inevitability of a bureaucratic and socialist form of
civilisation for the industrialized world, with much of the same fervour

Fivefold innovation

Schumpeter (1883-1950) identified five fields in which innovation occurs in the non-artistic sphere of human endeavour. These are, in paraphrase, as follows:

(1) The introduction of a new good (product, design, service)
(2) The creation of a new method of manufacture (process)
(3) The appearance of an innovative outlet (market) for either of the above
(4) Mastery of a new source of raw materials or of energy, and
(5) Conception of a new social institution (such as a firm or professional body), reflecting the emergence of any one or more of the four preceding factors

Austria of the Empire played a significant role in each of these innovatory practices.

---

that Saint-Simon had earlier predicted a glorious future for science, engineering and industrialization. Schumpeter, who became professor of economics at Harvard University, helped popularize the hypothesis—later reinforced by the Soviet economic theorist Kondratieff—of a cycle in the economies of the Western industrialized world. This notion seeks to silhouette the approximately fifty-year 'boom' and 'bust' peaks considered to characterize free-market capitalism. Relating the Kondratieff cycle to his own views, Schumpeter thought that the movements could be explained by the uneven emergence of the main innovations. (It is worth noting that the global economic recession that began early in the 1970s again made this long-wave theory fashionable.)[5]

Another economist espousing the theory of economic cycles is *Gottfried von Haberler* who, like Schumpeter, became a professor at Harvard. *Friedrich August von Hayek* shared the 1974 Nobel Prize in economic sciences with Gunnar Myrdal of Sweden for 'their penetrating analysis of the interdependence of economic, social and institutional phenomena.'

The Austrian 'revisionism' of Marx led to the founding of the country's own brand of social democracy, under *Karl Kautsky* and *Otto Bauer*. The influence of 'Red Vienna' in the 1920s and 1930s made the city a pioneering welfare community (*Gemeindebauten*), adorned in part by the Karl Marx Hof, a 1,400-apartment complex designed in the architecture of 'socialist realism' by a city architect called *Karl Ehn*. From 1919 to 1934, more than 60,000 dwelling units of the kind were built (90 percent of them in large apartment blocks); they were financed by a tax on private property devised by *Hugo Brietner*.

In philosophy, the name of *Ludwig Josef von Wittgenstein* is now part of our century's history. At first a mechanical engineering student at Berlin and Manchester, the young Viennese switched his interests early to Cambridge and mathematics (under George Moore and Bertrand Russell) and then to the logical atomism of philosophy. Wittgenstein saw the world as a collection of facts (*Tatsachen*), independent of each other, yet resoluble in the form of the 'state of things' (*Sachverhalt*). His thought descended from the lineage of Kant, Schopenhauer, Maxwell, Hertz and Mach, and his writings ranged widely. These derived to some degree from the experiential philosophy of the early phenomenologist, *Edmund Husserl* (a Moravian) and treated such abstractions as concept, meaning (including exactness and vagueness), certainty, reality and its doubts, thing and complexity, 'givenness,' ethics, will and religion, logic, understanding and 'world.' 'If I could describe the limit of the world,' wrote Wittgenstein in his *Notebooks, 1916-18*, 'then it would simply not be a limit. The limit of the world is not the limit of something extended. It is inherent in the world itself.' He wrote this as the centrifugation of the Austro-Hungarian empire was all but completed.

An era comes to its end

One of the last great innovations to occur in Austria before the historic events of the 1930s was the formation of the *Wiener Kreis* (Vienna Circle), a group of neo-positivist empirical thinkers who based their thought on the ideas of Mach and who flourished between 1921 and 1928. Interacting strongly with Wittgenstein, then in his thirties, the founder and pivotal figure of the Wiener Kreis was Moritz Schlick (a German who was to be murdered mistakenly in 1936). Schlick was the first to interpret Wittgenstein's *Tractatus Logico-philosophicus* of 1921, although Wittgenstein was to revise his own views, in this respect, after 1928. Schlick's notions of logical positivism were based on the distinction between empirical enunciation and the logical propositions of natural science—rejecting the 'pseudo-problems' posed by metaphysics—but embracing questions of ethical and aesthetic order. Other Austrians of the Wiener Kreis were *Viktor Kraft, Hans Hahn, Bela Juhos*, the above-mentioned Carl Menger and *Otto Neurath* (later a director of the *International Encyclopedia of Unified Science*, published in Chicago).

The Wiener Kreis would strongly orientate, in turn, what would become the influential thought of one of the twentieth century's leading philosophers, *Sir Karl Popper*. Paradoxically, Popper did not belong (and he was largely opposed) to the Vienna Circle. But the contributions of both to social conscience in contemporary knowledge have been as innovative as the fertility of Haydn's musical novelty two centuries ago, when the industrial revolution began and Austrian genius emerged.

Hindsight: some probable origins of innovation

Can some useful conclusions be drawn from this spectacular display of the creative spirit during the last seven or eight generations of the old Habsburg Empire? There seem to be several hypotheses to be made in attempting a *Kulturkritik* of a golden era.

First, the quality of training and education was consistently good during the nineteenth century, deriving from religious teaching and court-sponsored patronage of promising individuals. The 1848 Revolution and the transformation of Ringstrasse Vienna brought generalized public education to all the Imperial provinces (1869), so that each citizen completed eight years of schooling. As social progress spread, it nurtured individual talent for the arts and humanities as well as for the natural and engineering sciences.

Secondly, Austria-Hungary's geocultural situation surely played a significant role. An already far from homogeneous population welcomed newcomers, whether permanent or in transit (e.g., Metternich or Beethoven). The resulting mixture made Franz Joseph's capital a living intellectual laboratory and research centre. This geocultural interface continues to characterize today's Vienna, one should add.

Thirdly, and in curious contradiction to the previous feature, Austria regarded the world around from the time of the Congress of Vienna until Franz Joseph's death a century later through jingoist eyes. This attitude had consequences both good and bad on individual liberty and innovation throughout the population. There is an old theory that respect for human rights and a society's creative powers are correlated, although much work remains to be done in order to quantify the notions of personal freedom and the productive capacity of an entire nation. Yet, in the case of Austria,

the German-speaking middle classes accepted the pre-eminence of the feudal forces with their anti-scientific attitude. The reason for this readiness to subordination was that those middle classes felt threatened in their relatively favourable situation by Slavs and Latins.[6]

A Danube perhaps not so blue

by George Steiner

With signal exceptions, the noon and the darkness of the late Austro-Hungarian era, the incandescent energies and the crises of the Vienna-Prague-Budapest galaxy, are those of emancipated Judaism. To speak of the work and heritage of Freud or of Wittgenstein, of Kafka or of Broch, of Mahler or of Schönberg is to speak of the most creative and tragic chapter in the history of post-exilic Judaism. It is to recall the electric arc of spirit that connects the opening of the ghetto gates by the French Revolution and the Napoleonic Empire with the catastrophe of Nazism. And it was, of course, in Vienna itself, in Freud's city and Mahler's, that the first systematic and public program for the elimination of Jews from the life of Europe was devised and proclaimed. Hitler also is a child of that demonic caldron on the brown Danube. (It is in Budapest that the river is, infrequently, blue.)

Reproduced, by permission, from *The New Yorker* 28 January 1985.

So nearly two centuries of cross-cultural innovation occurred in spite of this situation or because of it.

Fourthly, we need take into consideration the interactions within Austria-Hungary between urban and rural populations. We have seen that simply educated and self-trained inventors from this country have given to the world objects and services now part of the everyday vocabulary in most languages. The phenomenon developed as agricultural efficiency increased and farms and rural villages lost their inhabitants to the growing main population centres. The Austro-Hungarian Empire was not spared such socioeconomic evolution; compared with many of its neighbors, the country seems to have benefited inordinately from such change.

Fifthly, a principal motive for the innovation we have reviewed may be attributable to what some have called a 'crisis of paranoia'—an extended state of group neurosis or even psychosis. The phenomenon is typified by seemingly comprehensible events (such as individual or collective persecution), and exemplified by the high productivity of Austria's Jewish population from the 1848 rebellions until the advent of Hitler. The country's Jews suffered from an anti-semitism which sometimes took virulent forms, so that it remains relatively difficult to understand the intense involvement of the same citizens in the experimental and adventurous causes taken up by the intelligentsia, and accomplished with such success.

The creative force, indeed the genius, having its roots in the Habsburg realms—whether artistic, technical or scientific—in turn gave rise to movements early in the twentieth century elsewhere: Dadaism, the Bauhaus, the Moscow and Paris schools of graphic arts and theatre. These gave way, in turn, to the beginnings of 'big science' meant to master nuclear energy, explore space, suppress epidemics and illiteracy, teach and feed most of 5,000 million people all over the planet. The historical legacy of the Austro-Hungarian peoples to make much of this possible has been astonishing.   ■

Notes

1. Beethoven was considered a radical and militant of the era, having dedicated his *Eroica* symphony to Napoleon as First Consul (a gesture withdrawn when Napoleon crowned himself emperor at Notre Dame, Paris). The characters Egmont and Fidelio were based, respectively, on Dutch nationalism and the oppressiveness of the Spanish monarchy. Both the Spaniards and the Dutch had known long relationships with Austria-Hungary.
2. Semmelweis's explanation: If young physicians, midwives and medical students would wash their hands thoroughly before examining a patient, the risk of infection would be minimized.
3. Johann Strauss's *Radetzkymarsch* is named after another son of the Empire, *Field Marshal Count Joseph Radetzky von Radetz*, a Bohemian who quashed the Italian revolution of 1848, then defeated the Sardinians at Custozza and Novara. It was Radetzky who brought the breaded veal *scaloppina* from Milan to his capital, where it became known as *Wiener Schnitzel*.
4. Schumpeter co-directed, with the Germans, Max Weber and Werner Sombart, the *Archiv für Sozialwissenschaft und Sozialpolitik*. Sombart linked the emergence of capitalism to Judaism, while Weber related the puritan morality of Calvinism to the economic rationalism which characterizes the capitalist system. It was Schumpeter who introduced the concept of innovation to economic studies of growth.
5. See G. Bianchi, G. Bruckmann, T. Vasko (eds.), *Background Material for a Meeting on Long Waves, Depression and Innovation*, Laxenburg, International Institute for Applied Systems Analysis, August 1983 (CP-83-44).
6. E. Broda, Ludwig Boltzmann, Albert Einstein und Franz-Josef, paper presented at the Convegno: Il ruolo della scienza nella letteratura austriaca nel secolo ventesimo, Trieste, 1981.

Bibliography

The literature available on the subject is ample, both native and foreign. Some of the most perceptive and explicative works, below, will guide the interested reader, who should also consult any of the many works of *Karl Kraus* (social commentator) and *Robert Musil* (author).

Bertelsmann Lexikon. Rheda, Bertelsmann, 1953.

BRAND, G. *Die grundlegenden Texte von Ludwig Wittgenstein.* Frankfurt am Main, Suhrkamp, Verlag, 1975.

Brockhaus Enzyklopdie. Wiesbaden, Brockhaus, 1966.

BRODA, E. Warum war es in Oesterreich um die Naturwissenschaft so schlect bestellt? *Wiener Geschichtsblätter,* Vol. 34, No. 3, 1979.

JANIK, A.; TOULMIN, S., *Wittgenstein's Vienna.* New York, Simon & Shuster, 1973.

JOHNSTON, W. *The Austrian Mind.* Berkeley, University of California Press, 1976 (2nd impression).

———. *Vienna, Vienna.* Milan, Mondadori, 1980; *Vienna, Vienna: The Golden Age, 1815-1914.* New York, Potter, 1981.

KNOLL, H. *Oesterreichische Naturforscher, Aerzte, Techniker,* Vienna, Verlag der Gessellschaft für Natur und Technik, 1957.

KREISSLER, F. (ed.), L'architecture autrichienne (theme). *Austriaca,* No. 12, May 1981.

LESER, N. *Das geistige Leben Wiens in der Zwischenkriegszeit* [The Intellectual Life of Vienna during the Inter-war Period]. (In press.)

MASON, J.W., *The Dissolution of the Austro-Hungarian Empire, 1867-1918.* London and New York, Longman, 1985.

Oesterreich Lexikon. Vienna, Oesterreich Lexikon, 1966.

SCHORSKE, C., *Fin-de-Siècle Vienna, Politics and Culture.* New York, Knopf, 1980; *Vienne fin-de-siècle,* Paris. Seuil, 1983.

SPAULDING, E.W., *The Quiet Invaders.* Vienna, Oesterr. Bundesverlag für Unterricht, Wissenschaft & Kunst, 1968.

STEINER, K. (ed.), *Modern Austria.* Palo Alto, Sposs Inc., 1981.

Türkenbote, Information und Unterhaltung (official guide). Vienna, Die Türken vor Wien (exhibition), June-September 1983.

Vienne: Fin-de-siècle et modernité (theme), *Cahiers du Musée National d'Art Moderne* (Paris), No. 14, 1984.

WILLETT, J., *The New Sobriety, 1917-1933, Art and Politics in the Weimar Period.* London, Thames & Hudson, 1978.

An eminent physicist, born and raised in the Austrian capital, records some boyhood recollections at the time of the disappearance of the Austro-Hungarian Empire. He describes some of the sources of originality leading to a spirit of creativity in many fields of culture—but to a much lesser extent in the natural sciences, because of the specific climate of Vienna before the Third Reich brought dramatic change to the ancient city developed by Marcus Aurelius.

Chapter 14

The spirit of Vienna: a testimonial

Victor F. Weisskopf

Victor Frederick Weisskopf is emeritus director of the Department of Physics, Massachusetts Institute of Technology, where he taught from the end of the Second World War until his retirement in 1974. A former director-general of the European Organisation for Nuclear Research (CERN) and past president of the American Physical Society, Professor Weisskopf is the author of many scientific works including Knowledge and Wonder, The Natural World as Man Knows It *(1979). He holds the Max Planck Medal (1956), the Ludwig Boltzmann Prize (1977), and the Republic of Austria's medal of honour (1982). Address: M.I.T., Department of Physics, Cambridge, MA 02139 (United States).*

I still remember the funeral procession of Franz-Josef in 1916, along the Ring, where my father had taken me. The structures of the Austro-Hungarian Empire were still intact during the First World War but, as though they had been undermined by termites, these structures were not to be able to survive the war; they would collapse. At the age of ten years, I happened to be with my father in front of the Parliament when the Republic was founded. There was civil strife, and the communists managed to tear away the white from the red, white and red flag—leaving two red flags hoisted. But that was the communists' only success. They had little influence then, and they have no more today.

The Austrian Republic was small, politically weak, and it was referred to as 'hydrocephalic' because Vienna contained a third of its population. The country was suffering from a kind of inferiority complex, the object of its own self-pity. Austrians went so far as not to believe in their country's survival, as witness by the poet, Robert Gilbert:

> *Mir war'm amal a Mordstrum Monarchie**
> (We were once a big hunk of monarchy)

Culturally, however, the new Austria was not as weak as that. So lively, so creative at the beginning of the century, the Austrian spirit had survived the war and was very much present.

It is often said that the creative spirit of the Viennese culture results from the intersection at Vienna of four cultures—German, Hungarian, Slav and Italian; and yet, I do not think that this is the real reason. Rather, Vienna represented a pole of attraction for all the enterprising spirit to be found in the 'provinces.' This is whither such a spirit led—that belonging to all those who wanted to get ahead and who sought success. They came to Vienna and adapted to the German culture with so much enthusiasm that they and their children became even more infused with German culture than the local people. Vienna was thus the gathering point for active, enterprising people from all over the Empire. This is why the city was such a cultural centre. Let me cite the names of Mahler, Schönberg and Berg in music, Schiele and Klimt in painting; Freud and Billroth in medicine; Musil, Schnitzler, Karl Kraus and Trakl in literature; Max Reinhardt in theatre; Mach, Wittgenstein and the Vienna Circle in philosophy; in social policy, there were Karl Kautsky, Victor and Friedrich Adler; and the architects Alfred Loos and Otto Wagner.

---

*In Viennese dialect—Ed.

The isolation of natural science

It is significant that a great many of these names are of Jewish origin, although they were in no way linked as such with Hebraic religious tradition. On the contrary, these families often came from the east; they assumed completely the German culture, as we have seen, and wanted to forget their Jewish roots. I felt this strongly within my own family. Although my father belonged to a small religious community originating in Bohemia, we children inherited little of the Judaic tradition. The essence of what we called 'culture' referred to a deep and emotional knowledge of the German legacy: that is, the works of Goethe, Schiller and Lessing, and the music of Bach, Mozart and Beethoven. We often visited, willingly, museums of art. We were inculcated with values that were, in a word, Victorian. My family was, in this respect, very representative of the cultivated class. At the same time, however, we were in constant touch with the cultural fermentation in opposition to the Victorian values that concerned contemporary music, psychoanalysis, Austrian Marxism, and modern art and architecture.

Conservative trends mixed with progressive ones can be found everywhere, but it is precisely such a characteristic of Austrian culture than an ambivalence can be found in one and the same person. One can find among the same individuals or even groups traditional tendencies, Victorian ones, their rejection, and the appeal of novelty. There is no lack of such examples. Schönberg expressed, in one of his letters, his sympathies for the monarchy. At the end of his extremely modern *Pierrot Lunaire*, the composer reverts to simple harmonies in order to describe Pierrot's return to this homeland. And the highly penetrating critic of his times that Karl Kraus was never concealed his admiration for the classics. Freud, despite the profoundly anti-Victorian nature of his theories, had very conservative ideas on art and he lived in one of the most traditional of apartments. The words of Musil are an amalgam of ironic social criticism and love for the heritage from the past.

Here I would like to make a comment that may perhaps offend some people. The cultural climate of Austria was scarcely favourable to the growth of the natural sciences. One can cite, of course, Ludwig Boltzmann—but he was the exception that confirms the rule. Most of the great names in the world of natural science shifted their activities abroad, and this well before the Nazi period. Some of the more celebrated cases involved Paul Ehrenfest, Wolfgang Pauli, Kurt Gödel and Erwin Schrödinger. (Ernst Mach and the Vienna school were philosophers, without being specifically scientists.) When I spent the first two years at the University of Vienna, I had the opportunity to take the courses in basic and theoretical physics taught by Hans Thirring. He was an admirable teacher, with a particularly rich and generous personality. He is remembered for the way in which he fought for peace and reconciliation among nations. And although he was a serious scientist, he was not of the first rank. After two years of study, I asked him if I should continue. He replied, 'If you really want to get into modern physics, leave Vienna as soon as you can and go to Göttingen or Munich, where you will be in the midst of today's issues.' I followed his advice and went to Göttingen.

The role of the human element

Why was the Austrian atmosphere unfavorable to the natural sciences? This is not easy to explain, but let me try to do so.

The strength of natural science lies in its not raising the 'fundamental questions;' instead, it puts these aside. It goes into detail and raises the question of *how* rather than that of *why*. It deals with measurable data, which means that a good number of the human and philosophic aspects are avoided. C.F. von Weizsäcker has explained this as follows: 'Philosophy raises questions that, for the scientific process to succeed, should not be raised. Science owes its success especially to the fact that it is reluctant to bring up certain questions.'

With this in mind, the human element figures very prominently in the spirit of Vienna. It seeks accommodations, it sees things in the most varied lights. As a consequence, the Viennese attitude is occasionally superficial, it sometimes is satisfied with a general idea or settles for what the Viennese call *Schmus* (gossip) or in café talk. Whatever might be the case, the human element has an important role to play, one that ignores science.

I am also, in my own way, heir to this situation. I owe to my Austrian origins that I have been able to save myself from pure specialization and from being nothing but a specialist. I prefer to know nothing about everything to knowing everything about nothing.

The period following the First World War was still a brilliant one for the Austrian spirit, despite the fact that the country had lost its political might, and this lasted until 1934. That was when a bloody civil war broke out between socialists and conservatives. Then, in 1938, came the nightfall of the Nazi period—times we would like to forget. But things are not as simple as that, when we reflect that a great majority of the Austrian people followed the Nazis.

This dark interlude lasted for seven years, instead of the thousand that we had been promised. A neutral Austria finally emerged, and this neutrality turned the country into the real heart of Europe. The unexpected came true. The 'water on the brain' was forgotten, as were any doubts entertained earlier about the little country's viability. Its industry and trade prospered, and Austria became a model of modern, effective and humane government. Austria once again brought forth a creative idea: a social contract between government, the labour unions and industry. The socialist and conservative parties worked hand in hand to make the contract work. And the country that (after 1945) had lost much of its cultural influence became, as a result of this social experiment, a new example of the cooperation between modernism and tradition that had been one of the constant features of the Empire. One of the ensuing consequences has been that the impact of the economic crisis of our times has been less harsh than elsewhere.

I would like to end with an observation to show why Austria is, to me, so pleasant. The cumulative budget for the Opera of Vienna, the National Theatre and the subsidies paid to private theatres is almost as great as the military budget.

Note:
The author presented the above in the form of a talk, on the occasion of his seventy-fifth birthday, at the Duino (Italy) Conference of 1983 where the theme was 'The Central European Presence.' Reproduced with the kind permission of *Cadmos* (Geneva).

PART IV

Creative Process

A leading design and research engineer in 'high-tech' industry discusses the overall R&D process, the sources of inspiration and execution of novel designs and products, the significance of sharing patents, the merits and shortcomings of today's technological wizards, how they are managed and compensated and honoured, and how the new generation of innovators is accommodated in the ranks of the elders.

Chapter 15

Modern inventors in high-technology industry

Interview with Guy LeParquier

Technical director of the Equipment and Systems branch of the Groupe Thomson, Guy LeParquier received on-the-job training in the electronics industry then was graduated as an engineer at the Conservatoire National des Arts et Métiers in Paris. He has worked in advanced technologies for more than thirty years. Address: Directeur Technique, Thomson-CSF - BES, 23 rue de Courcelles, 75362 Paris Cédex 08 (France). This interview took place in summer 1985.

What brought you to the world of technical innovation and creativity?

I would normally have been due for military service in 1944, when I was twenty. But the year before, I received orders to report for obligatory labour in Germany. I was able to be deferred for scholastic reasons and, by the time the war was over, I had done no military service at all. Despite this, I was fortunate to obtain employment in the radar field when I was hired by the Sadir firm.

Did you have a good idea of what you wanted to do in life?

Oh yes; I wanted to learn science, but having come from a modest background, I took vocational training and then went to work at the age of 18 for Alsthom in order to earn a living. (It was while I was at Alsthom that the business about service in Germany came up.) I began taking courses at the Conservatoire National d'Arts et Métiers (a large engineering school in Paris). After the Liberation, I left Alsthom's employ and continued my studies. So there I was lucky: if I had been packed off to Germany, I would probably never have gone into higher learning nor ended up in radar.

What I learned from this early experience is that when I made my own decisions concerning a new departure in life, it was up to me to follow through. My lucky star, therefore, has guided me to do things in which I really believed—rather than trying to make good on something in which I had no faith. As a consequence, I seem to have built up an 'optimism potential' that I have made it a point to exploit to the hilt. So it seems to me that the formula for success includes being at the right place at the right time and having a good leader; this has been the case with me ever since I began working with radar.

At 'Arts et Métiers' I undertook studies in physics. I began employment at CSF which gave me the title of engineer before I had the degree; and afterwards I became laboratory director, division director, technical director of the radar department, and eventually director of this department.

Have you been concerned since with conception and production of with meeting the needs of the consumer market?

I was first responsible for radar research and work on prototypes, and during this time I adopted a functional view of what was being manufactured, so that we were not turning out material unrelated to our research effort. (This can happen when an organization is too compartmentalized.)

Then, when CSF and Thomson merged, I was transferred to the General Technical Department to supervise the merger of the two radar groups in 1968-9. This meant a restructuration permitting everyone to carry his own weight—to find a kind of organizational synergy, combining two avionics teams with ground and ship radar units. Then I was asked to oversee the merger and preparations for future activity of other technical fields, including telecommunications and armaments. By 1974, I had been named chief engineer of the group responsible for technical co-ordination of overall activity, including consumer goods, medical engineering and so forth. This is a staff function rather than one of line management, in which I serve as a counsellor responsible for cross-fertilization between specialists.

You did not, in other words, command troops; you were a general staff officer.

Precisely. This I continued to do until 1983, prompting bilateral contacts between individual researchers and organizing seminars of general interest to large groups of our investigators. Another function that I have now is to be (theoretically, at least) fully conversant with everything going on throughout our laboratories. In fact, I am perhaps 60 percent informed about this, sufficiently however to enable me to prepare summaries that lead to decision making.

You deal with how many research personnel?

There are about 15,000 of them, spread among fifty different centres, both in France and at Willingen, Bremen and Koblentz in the Federal Republic of Germany. Not long ago, our company acquired the German brand-names of Telefunken, Dual, Saba and Nordmende. I have worked closely with Pierre Aigrain, the well known solid-state physicist, and his successor, Michel-Henri Carpentier.* My role has been technical co-ordination among all the disciplines represented by our research staff, whereas Aigrain and Carpentier have been responsible for managing our Central Laboratory (staff of 400), our Patents Department, our Proprietary Department (of about 65 persons), and our Technical Information Department.

---

* M.-H. Carpentier is a graduate of France's Ecole Polytechnique and aeronautical and telecommunications engineering institutes.—Ed.

So without being a decision maker yourself, you prompted those who took the decisions?

Yes, and I have sometimes prompted rather loudly. Since nationalization in 1983, our umbrella organization has been properly structured with complete corporations, each corporation having its own financial, management, accounting, manpower, planning and technical staffs. Each corporation is headed by a delegated managing director (*directeur délégué*) reporting separately to Alain Gomez as chief executive officer. Each delegated director is also responsible for total staffs of between 15,000 and 36,000 employees concerned severally with Thomson-CSF's activities in consumer products, medical material, components, military-aeronautical, industrial, and weapon-system areas. There is an overall technical director, but the bulk of the research and development is carried on within the corporate companies. I have been responsible since 1983 for the *Direction technique* of our military-aeronautical set-up.

How do you co-ordinate with the planners?

The chief planning manager sits across the corridor from me. His role is to make (a) short-term forecasts with reasonable certitude, and (b) mid-term forecasts with a good level of probability based on foreseen or planned market development. Avoiding the obvious pitfall of simple extrapolation from the past and present is tricky: simple yet complex. The planning function requires that you, as the planner, have access to anyone you wish in the organization. I am lucky in the respect that people tell me everything—what they are doing, what they are preparing, their feelings, their desire for change. This is, in other words, *confidence*, built up over the years. Planning also requires the planner to be ready to help others—to be ready to invest time and effort in the resolution of others' problems. In my case, I help to resolve individual problems, whether technical or human, but I must deal with everyone so as not to become selectively obstructionist.

The normal managerial hierarchy is so committed to finding short-term solutions to immediate problems that it cannot take the time to be up-to-date on everything that happens in the house. Whenever I am involved in an encounter with a colleague where I sense some delicateness, I inevitably ask him or her, 'What should I keep to myself, and what do you wish me to follow up?' The ensuing arrangement is mutually respected.

Does this building of mutual respect stem from the fact that you are an inventor, that you hold patents?

No, I don't think so. The way to make this sort of mutual respect foolproof is by way of what I call legend. Let us say that I am responsible for about five one-thousandths of all French inventions being produced today. I do not keep these patents for myself; I take them out together with him or those who find applications for them. It takes at least two parties to innovate. I mention a novel idea to a colleague, adding, 'It's yours; take it.' He retorts, 'No, it is you, the inventor.' I respond, 'Possibly, but the invention is really ours: if I keep the idea to myself, it dies—nothing will come of it. If you will follow through, even abroad, then we share the invention.' This is how the patent process should always work.

Should work, perhaps, but not always so. Edison was famous for arrogating others' inventive efforts to his own credit.

True, but an individual's behaviour corresponds to his own moral make-up. I personally need to be able to work in a relaxed atmosphere that permits a totally free exchange of ideas. I know that if I let myself be motivated by the thought of another patent to my name, then I must be on the wrong track. As a matter of fact, if I allowed myself to work this way I would probably garner only a patent a year, because I would not have the necessary raw material for inventive work. The patent cannot be a goal; instead, it should make abstractions concrete and protect the innovator.

Does your attitude result from corporate influence, or is this a personal matter?

The ideas are mine—a personal matter.

A few firms permit their research staffs to conduct a limited amount of personal research at corporate expense, on company time. The 3M group is famous for this. How do you react to such a policy?

The question is perhaps even more general, and a more subtle one, than you suggest. The policy probably depends on the industrial research sector involved.

Our Corbeville Central Laboratory has about 200 researchers and it enjoys great investigative freedom. Fifty percent of the ideas they produce come from individual initiative, and the rest from the group as a whole. But the research done should, nonetheless, be marketable later by the Thomson group. If you will remember that basic research is not our mission, I can say whereas 80 or perhaps 90 percent of our research next year will be based on goal-orientation, our plans are such that the remaining 10-20 percent will emerge and be developed from unforeseen initiative. In product R&D concerned with materials and components, there is less freedom of research, yet those in charge must maintain a small margin of latitude. In the case of production development, the only margin left to the engineers is that required for product improvement or changes in methods of improving the product.

How does your firm handle the payment of premiums for patents?

The filing of any patent, whether very good or otherwise, brings a minor premium of several thousand francs, to be shared when there is more than one inventor. Then each organizational unit can award a more considerable bonus based on the value of the patent and how the unit wishes specifically to reward its inventors. Finally, each year there is a competition to judge the best patents filed during the preceding five years.

The definition for the last being, presumably, which inventors have earned the most for the firm?

Yes, *have earned* the most, or *will be earning* the most, over a five-year period. You can imagine that this is sometimes difficult to estimate. In certain cases, the decision is deferred if a patent holds significant promise of return that is not yet reflected in the annual turnover. Profits from patents, moreover, differ—depending upon whether we wish to guard against counterfeiting, negotiate exchange deals with other firms, or license manufacturing, especially in the Third World. Patent extension abroad, and specifically in which countries, needs to be decided during the first year since a patent expires after 20 years. The cost/benefit return on a patent can be difficult to quantify because patents are often used to discourage litigation.

Some patents have not earned, officially speaking, a cent because (a) they were never licensed, or (b) there was never a follow-up of production, but (c) litigation was avoided because it was possible to negotiate an exchange with another industrial enterprise.

Do you deal with free-lance inventors?

Yes, but fairly rarely.

There are independent inventors from outside the firm who work with us, for example, in the case of some of the work done in universities. There are no problems in such cases.

In the cases of patents assigned to other industries (our competitors), occasionally we arrange exchange agreements. And then there are the patents assigned to inventors, especially inventors brought to our attention by ANVAR (the French *Agence Nationale pour la Valorisation*, the agency dealing with venture capital and promotion of the innovation processes). A firm such as ours can show the well-known 'not invented here' syndrome, whereby there is a tendency to disregard anything new coming from outside.

It must be said honestly, however, that intermediaries tend to send duplicative ideas on innovation; or else they make proposals outside our fields of interest; or else the ideas already exist in the patent literature. And the reason for all this is quite simple. It is very difficult to invent something—something original that works—if one is not already completely absorbed in the topical area concerned.

You mean that there is much 're-invention'?

Yes, and there is little mystery about this because among ourselves three of every four 'new ideas' that we have can be eliminated after a simple search of the literature.

How stable is your staff of innovative minds?

Our personnel stay for quite a long time. Statistically speaking, creative people tend to move about less than many others because they are inventing in their own field, they are comfortable in their chosen life, and this is the way they wish to continue.

As I see it, the creative spirit is the beneficial aspect of the person who tends to contest things. People who do not contest extrapolate the future on the basis of the past and the present, thinking that nothing needs to be questioned. To create something, I believe that one needs constantly to ask if the way we do something now is justified by everything else that has changed.

The technical solution to a new problem is usually the one that can be implemented by using existing technology. Often, five to 10 years later, new technologies evolve and one can never be sure that these might not put to serious test an entire system of doing something—as opposed to being simply new ways of doing the same old thing. Indeed, our natural tendency seems to be to extrapolate and then use a new technology to do exactly as we have been doing. So it is always essential to ask: Will not a new technology allow us to do things differently?

But isn't this another way of discouraging simple extrapolation?

Yes, because sometimes one is led to wonder about the purpose of a new solution being sought, or even what is the real problem that has to be solved. Numeric technologies—memories and microprocessors and the like—for example, have provided considerable degrees of freedom in the conception of new equipment. Their functioning can be different: more complex but more stable, and not merely the expression in numeric terms of what had been in analogic ones.

It is a question of developing an entirely different cast of mind. I can express this way of thinking by another example, in an entirely different field, and almost to the point of joking: Without spending a single cent, how could the payload of a commercial aircraft be raised 2 percent?

Perhaps by charging more to carry more valuable cargo?

No. It would be a case of adding, in case of the A300 Airbus, six additional passengers without changing a thing—not even the configuration. My solution is very simple, and this is to make the duty-free shops available at the end of the flight instead of at the departure. Each passenger carries about 2 kilos (two bottles of whisky), so the weight saved would equal that of passengers occupying six additional seats. Duty-free economic operations would not suffer because, statistically, every passenger makes a round-trip flight.

I would not have thought of that. How do such ideas come to you?

Well, a few conditions are requisite. First, all has to be well with the world; in my case, all *is* well with the world. Worries of whatever kind—health, family, professional, sentimental—can impair creativity. I know because I have experienced successively blank periods of six months and one year when I was not creative at all. One must not have any kind of problem . . . .

A second condition is to have a maximum of knowledge, as multidisciplinary as possible. Invention is a matter of cross-fertilization between different elements of knowledge, and an exercise of combination of knowledge is possible only with what we know. Those possessing a limited data bank are capable of correlating relatively few things. So, one must be curious about everything.

A third condition, already alluded to, is always to re-ask questions in order to understand why something is the way it is. Is the reason traditional or justified?

A fourth condition is that one not fear ridicule, for when one gets an idea there are nine chances out of ten that it is not good or that someone else has already thought of it. I always make the point that the harvest rate of new ideas is very low. So one must have many new ideas and not be vexed when they are poor. Many people experience this kind of blockage.

And I see that you make frequent, short notes, in an extremely fine hand.

Yes. Now, how is an 'invention atmosphere' arranged?

First, I try to arrange meetings that are relaxed and without hierarchical order. Everyone has the right to say what he or she wishes, and I set things in motion by giving a few examples of bright ideas. My own conviction of the validity of this approach was confirmed during some creative brainstorming courses that I took at the firm called Cinectics. Everyone must carry a stone to the other's construction site when he has a new idea and not say that idea will not work, for whatever reason. In this way, little by little, a supportive atmosphere is developed and the construction proceeds—after which certain decisions need to be made.

Who makes the decisions?

Those capable of following the idea through, industrially and commercially speaking. At each meeting, one of the participants prepares a report (or, at least, a list of all the ideas evoked). Then someone is designated, not necessarily possessing hierarchical rank, to winnow out the best new ideas that seem worthy of exploitation. He (or she) submits the report to his supervisor, and provides me with a copy. Then the three of us sort further the ideas to be pushed or to be investigated further. (Here I act in my consultative capacity).

But it is up to the division technical directors to decide if they wish to proceed with the innovative idea, whether they will invest money, because often pre-analyses are required: a technical study and a market analysis.

These feasibility studies are essential before committing substantial sums on either the technical or the marketing side. This can be as difficult a process as the product is original. A market study on an existing product or its improvements is relatively easy, but a study on something not yet in existence is very difficult because it can depend on (a) how a question is asked of a sample group, or (b) how the new services to be offered are described.

Does your own staff propose in-depth market studies?

Sometimes the proposals come from outside. Each division has its own way of doing things, but always in consultation with the proper technical staff.

Do not your comments imply that, too often in technical creativity, insufficient attention is paid to market potential?

This is true, but the fundamental problem is that marketing specialists cannot know what questions to ask when a product or service is not yet in existence.

I have mentioned the creativity sessions in which I am involved. More usually, I submit ideas to our engineers when I visit them so that, in turn, new ideas can be generated.

So creativity really begins inside the house. How much of your time would you say is devoted to the creative processes?

Easily three-fourths. I am employed not only to originate things, but to guide and advise others doing the same thing. In the development of new weapon systems, we have to think ahead ten to fifteen years—but without being strictly linear in our thinking. We need to repeat to ourselves that changes are constantly necessary: the coming of lasers, of hyperfrequency integrated circuits will require the replacement of this and that. It is by thinking of these and many other things that new perceptions develop.

In conceiving new weapon systems, I imagine that you need to be one or two steps ahead of your own invention in order to envisage the countermeasures that could be used against your own devices.

As far as my own work is concerned, I develop the new system and its countermeasures simultaneously. Countermeasures are an important part of the work of this house.

In visualizing new generations of successful products, we have tried on two occasions to dialogue directly with the public along a theme that goes like this: 'Assuming that technology is capable of anything, what improvements could be made to television receivers and what would this be worth commercially?' Stated otherwise, how much would the client be willing to pay for the improvement he seeks? The point is *not* to make improvements first, then *see later* if these can be sold.

How do you envisage new lines of activity that are barely related to Thomson-CSF's current research and production?

It is a three-fold problem. The first step is a preliminary study to define new domains unfolding; here, the commercial (sales) volume is most difficult to estimate. Second step: we begin to construct a prototype of the product (design, process, service). At this stage, financial commitment is more critical than in the first stage, and we can make the commitment only if the probability of success is sufficiently high. Final step: the industrial realization itself, when the product (or other) comes into being.

The relative proportions of financial commitment for the successive steps are roughly 1, 10 and 100. We can obligate the last value, as I have said, only if certain conditions apply, and these can be multiple. First, the product needs to be able to work. Second, the product must be attractive, it must draw a public (market); when competitive items exist, our own cannot lag far behind. The goal is to be first among several. Third, there must be a constant redefinition of the market and the place of our product in this market; this can be critical if our product is somewhat isolated from the rest of what we manufacture and sell.

To succeed, in other words, one needs competence in technical conception, manufacturing ability, and sales achievement. If one of these capacities is lacking, success is doubtful. With two of them wanting, failure is sure to follow. So one of my responsibilities is to create competence in areas where money will not be a problem.

Thus, part of the ingenuity is to steer technology to 'where the money is.' Steering technology implies continuity, in your own efforts and in the work of the successive generations of persons such as you. How do you see the new generation in this respect?

As in the case of just about everything else, the new generation has its good points and its bad: from the passionately committed to the don't-give-a-damn types. There are those who work with fervour, and those who work away quietly at their projects.

The significant change seems to be in the pecking order, something that I sense very much. Previously the engineer was considered one of the bosses: he was mentioned in the same breath with the executives. The engineer had a supervisory function—he commanded troops. But as design and manufacturing problems grew increasingly complex, more and more engineers were hired, to the point that today engineers are no longer executives; they are implementers, executive agents.

Is this because they are salaried?

No, the executives are salaried too. The engineer has lost his function of giving orders. It used to be (here in France) that, once out of engineering school, the young graduate entered on-the-job training for six months then was named a section head. Now the graduates work as technologists for years on end, without having a managerial role; and despite their training as engineers, they are no more motivated than qualified fitters unless the job truly interests them.

Yet the engineer can look forward to a better career than, say, a model-maker.

Of course, but the average engineer is no longer likely to become president of the firm. For most technical personnel, yearly raises in compensation in order to offset inflation have been the rule for some time. More recently, we have begun granting supplements annually to compensate not only for inflation but for personal improvement as well. The difference now is that the raises are no longer automatic. Each employee is called by his *supervisor's* supervisor at least once a year to review performance and to discuss the increase itself. The boss's boss is thus obliged to ask: What does this employee do? What is his potential, and what kind of future has he in store?

Do you have women employees?

About 10 percent. Recently, women have been especially attracted to work involving data processing. I refer to women engineers, of course.

But they work in teams, together with men?

All our activity has become teamwork. The resources needed today to bring an invention to fruition require that people work in teams, which in turn puts pressure on supervisors and managers to know how to bring people together and make their combined effort pay off. The management of humans is, in the final analysis, the one that counts most. Even a fairly homogeneous team of only mediocre capability, technically speaking, will not bring forth a new leader. If the homogeneity reflects, on the contrary, high potential on the part of everyone, then the whole team fights all the time. The solution is a fairly heterogeneous group among which there are one or two persons with leadership potential. But here, there must be no demotivating, no destabilizing factor.

Often a leader emerges almost automatically. The team considers him or her the chief. After this leader has been able to prove himself for a while, that is when management can best award title and pay. ▮

Besides the frequent interpretation of the creative spirit as endemic to the world of arts, letters and social progress, innovation is the keystone in the structure of advancing industry and commerce. Manufacture and trade depend, in order to be competitive and thus successful, on the motivations and impulsions inherent in the creative process. These are discussed by a teacher in a leading school of management.

Chapter 16

The challenge of managing inventiveness

John J. Kao

Dr. Kao is a physician who holds other advanced degrees, including a master's in business administration. He teaches organisational and managerial behaviour to graduate students at Harvard University's Graduate School of Business Administration; he also conducts research in entrepreneurship and general corporate activity, and he is a consultant to technology-based industry. His avocational interests are media production, jazz piano and high-fidelity musical reproduction. Address: Professor J.J. Kao, Harvard Business School, Soldiers' Field Road, Boston, MA 02163 (United States)

The creativity-entrepreneurship link

To speak of managing creativity may, at the outset, suggest a paradox. For surely creativity springs from the innermost recesses of human vision, the most intimate reaches of our experience. Creativity is thought by many to consist of inspiration, hence its comings and goings are thought to be correspondingly unpredictable and personal. And the very term, management, seems opposed to creativity, suggesting instead the values of the collective, the need to balance diverse interests, the need to expose creativity to the harsh light of planning.

Yet another difficulty presents itself. Creativity is commonly thought of as a quality which adheres to artistic endeavour and to the expression of cultural experience. Artists are by definition creative, or so it is thought. Something about their talent is innate; they are born, not made. In this way we might feel ourselves spared from having to examine creativity more closely and from having to consider whether something as elusive, as intangible as creativity can actually be *managed*.

Far from being germane only to the arts, creativity is a resource important to the basic institutions of society: industrial and business organisations, educational institutions, social and community agencies. Indeed it may be argued that creativity is latent in many collective situations, and that the extent to which it is fully expressed determines whether a group reaches its goals and the fulfillment of its destiny. Creative and organisational processes must engage, must mesh with each other if social institutions are to be fully productive. Seen in this way, creativity is not something special, for special people in special situations; it belongs to everyone.

To consider creativity in terms of cultural refinements can be a sterile, even elitist exercise. Spoken of in isolation, creativity becomes simply another form of social 'good', a notion not to be argued with, a platitude. Like 'excellence' and 'productivity', creativeness can easily be reduced to the level of lip service, something of which we could all use more, without feeling obligated to subject it to closer inspection. This is a shame, because creativity can and indeed must be managed. By management, we mean conscious planning and assessment in the interest of practical decisions shaping the context of human activity. There is no such thing as creative spirit divorced from a particular human group, from specific sets of social institutions and values, even from a given set of political constraints. Creativity for whom is as important as the value-laden issues raised by the question of creativity for what.

It may help at this point to offer some working definitions of creativity and the related term, innovation, which have helped me in previous research efforts:

Creativity may be thought of as a quality or talent leading to a result which is novel (new), useful (solves an existing problem or satisfies an existing need), and understandable (can be reproduced). The process involves interplay among a person, a task and a social environment.

Innovation suggests the process of implementation by which creative inspiration leads to practical results.

Creativity is strongly correlated with yet a third term, that of entrepreneurship. We might think of entrepreneurship defined as a responsiveness to opportunity and by the freedom (in both personal and collective terms) to act on this opportunity. It is perhaps in this last term that we find the notion of implementation, of doing, most firmly stated. If creativity implies a vision of what is possible, the entrepreneur translates his or her creative vision into action, into a human vision which guides the work of a group of people. In this way, creativity and entrepreneurship are linked. There is also a good deal of leadership which may be required.

Creation as both art and social experience

How can we learn more about creativeness? How conscious are we of it? Can it be taught? It has been argued that everyone has tremendous potential to be creative, and much effort has gone into training that enhances creativity and allows it to emerge, even by group training. The individual's ability to be creative, moreover, is affected by a number of environmental elements which can be managed in order to foster creative results.

We may learn more about the creative spirit in a particular organization by understanding the obstacles to its expression. What gets in its way? What helps it? What blocks the upward flow of ideas within an organisation? Common sense tells us of some obvious hindrances: bureaucracy, the absence of suitable role-models, needed resources and values supporting creativity such as the freedom to fail or to question.

In the course of our research at the Harvard Business School, we have examined creativity as a dimension of planning and social investment. We have reviewed the experiences of a group of organisations, ranging across the spectrum of private and public activities. It can be argued that even governments and international agencies such as the United Nations or Unesco are environments in which the management of creativity is important. And in 'creativity-intensive' industries such as biotechnology, computer programming, media production and so on, the need to manage creativity is even more obvious.

Managing creativity in institutions leads to a number of classical dilemmas. Can creativity on demand exist? How does one balance the need for freedom inside an organisation with the need for structure? How can the creative person's desire for unlimited development be balanced with the manager's need for deadlines? How does one reconcile the creative need for participation and informal communication with management's need for hierarchy and structured authority in the more complex organisations?

The creative spirit and management

Managing creativity also includes a host of issues involving the creators: scientists, artists, engineers, computer programmers. Are these different from other people? Do they need special treatment? How will such persons be identified, recruited, trained, developed, motivated, evaluated and compensated? What do they need in order to be most creative? How many such specialists are needed, and how much will they cost their organisation? What institutional system will assure that creative people work together in the most productive and appropriate way? How are jobs designed to take best advantage of creative talents? What styles of leadership and role-models best support an organisation's creative needs? If predictable conflicts arise between the managerial and creative sides of an organisation, how can these best be managed? To what extent is there value for the organisation in such conflicts? Many further questions could be raised; my purpose here is simply to offer a sample of the issues.

Why is the management of creativity important? First of all, new types of organisational structures are required to deal with society's need for going beyond current limits. Secondly, creativity is a resource that requires nurturing if it is not to be squandered. Some young organisations are, in a sense, laboratories in which the organisational structures of the future are being developed—managing explosive growth, the impact on the organisation of women's rights and of the greater diversity which comes with growing social equity, the need to create rapidly a culture and a set of values to guide creative activity, and the challenge of maintaining fairness within a creativity-intensive organization. Managing the kind of uncertainty associated with intensively creative organisations also requires a particular set of skills.

Managing creativity for the grand-scale goals

Creativity is thus on many minds; it is at the heart of much work being done on industrial policy at national and local levels. Creativity is furthermore, and perhaps more importantly, a matter for those involved in fostering innovation.

A new kind of organisation for managing creativity, for example, has come into being during the 1980s which I call the entrepreneurial conglomerate. The Japanese project dealing with the fifth generation of computers, the Centre Mondial de l'Informatique in France, and firms such as Catalyst Technologies in the United States are examples of this trend; small and medium enterprises in many countries, even in China and Hungary—are scaled-down versions of the same.

This is an era in which the management of intangibles in organizations is increasingly important. Creative thought is only one example of such intangibles, and it is in this vein that much of our research and course development activities at the Harvard Business School have been carried out. A new course (see box) in the human issues related to

---

Teaching human issues to the entrepreneur

The Harvard Business School's course on Entrepreneurship, Creativity and Organisation is based on two assumptions: first, that human and institutional issues are as important to entrepreneurial success as the 'traditional' business concerns of finance, marketing and the ability to define new opportunities; secondly, that fresh insights on the relationship between organisations, strategies and environments can result from careful examination of entrepreneurial firms.

The course began in Spring 1985 and is divided into seven segments: the entrepreneur, partnerships and teams, managing creative resources, project team management, human resources' management in the entrepreneurial firm, corporate entrepreneurship, and the entrepreneurial process.

Using the case-study method, it is to be noted that one of the cases studied in the category of 'partnerships and teams' is The Beatles.

— The Editor

---

entrepreneurship and creativity made its debut here in 1983. It is also clear that the creative urge is at the heart of many societal goals—whether these are to find new ways of training citizens, to encourage the development of scientific activity, to maintain the human environment and its resources, or to find new ways to manage conflict and improve the quality of life.

It is in these more global senses that the management of creativity will find its fullest expression. ■

See also:

By the same author, *Entrepreneurship, Creativity, and Organization*. New York, Prentice Hall, 1988.

The real-life problems faced by people and by nations have become enormously complicated. To consider these as issues independent of social habits and psychological understanding is inadequate. The classical approaches taken a generation earlier to deal with food, energy and transport as challenges to society no longer succeed. Now, using system theory, a problem is approached from many different points of view to find connections between a given problem and those related. In system applications, we proceed from the theoretical definition of a problem to concrete ways of solving it.

Chapter 17

Using the social and natural sciences to solve large-scale problems

Interview with Djordjija Petkovski

Dr. Petkovski, a system engineer, is on the staff of the University of Novi Sad where he teaches automatic control and computer techniques. Since completing his post-doctoral studies as a research fellow at MIT, he has been especially concerned with the application of systemic methods to the resolution of large-scale problems peculiar to developing countries such as his native Yugoslavia. The author is Director, Centre for Large-Scale Control and Decision Systems, Institute for Measurement and Control, University of Novi Sad, Veljka Vlahovića 3, 21000 Novi Sad (Yugoslavia), telephone (021) 55-622. The cartoons are by Richard Erickson, a Canadian artist working in Paris who uses an Apple Macintosh to help him with his work.

What was the genesis of system theory, and how are you making use of its fundamentals?

We cannot talk about the history of systems without referring to the German philosopher and mathematician, G.W. Leibniz (1646-1716), who was also a geologist, chemist, historian, diplomat, politician and theologian: he sought the integration of all the disciplines. In our time, this trend was reinforced by Ludwig von Bertalanffy,* who published his ideas on system integration for the first time in the 1940s.

Thus, when system theory developed, roughly at the time of the Second World War, it was considered to be a philosophical view—often expressed by an overuse of mathematics. Both theory and applications have progressed very much since the coming of 'big science', and some of the mathematics has been replaced by a clear and logical approach to how we teach system theory and how we put the theory into practice.

At Novi Sad, we start by using simple, small-scale examples: first-year students can appreciate quite readily the interconnectedness of simple problems. And, of course, the same method works with primary and secondary level students. If teachers can encourage young students to think logically about everyday problems, this is in itself an application of system theory.

---

*See also Chapter 13.

In other words, 'doing' systems without realizing it?

Yes. If I want to learn swimming or skiing, I can spend the next twenty years trying to teach myself without ever really learning. The alternative is to learn from an instructor, whose method summarizes generations of experience in teaching systematically. The same systematic method can help us understand what is meant by system—learning the systemic approach towards almost any kind of problem-solving.

Do you require that your students have a fairly good mathematical background?

Yes. They join us at the age of eighteen and remain for about five years. Upon graduation, the students receive a university diploma in electrical or mechanical engineering. The students' first year is spent in general studies. The students are introduced to system theory only in the second year. This is what is called automatic control or cybernetics in other countries; it deals increasingly with large-scale networks, such as a country's electric power-grid. Complicated technological processes include calculation of the losses of current in an electrical distribution network, operations in petroleum refineries, the supply of raw materials to a factory and then the distribution of its products. The final step is to proceed to larger economic and other social systems, block-diagramming their main components to show how these are linked and how they affect each other.

I should emphasize that the theory and applications of large-scale systems are rather new even for developed countries. And when we talk about the technological gap between industrialized and developing nations, the gap is particularly noticeable in the cases of large systems and computer technology. Overcoming this gap—trying to close it—was one of our intentions when we began working with Unesco: to help developing countries, to help our own country. We are concentrating on problems fairly specific to an economy still in the process of industrialization, such as the food-processing industries, as these can be instrumental in generating badly needed hard currency. Our need is, in a word, innovation.

How indispensable is the computer to someone learning system theory and application?

The rapid development of computer technology has helped the development of large-scale system work. The computer helps us to make complex and duplicatable models. Some people believe that a computer can solve any kind of problem, an expectation that can have a negative influence on the evolution of new techniques in our field. The computer is merely a tool for problem-solving, but in developing countries the shortage of computers is matched only by the lack of trained personnel to operate them. One of the main goals of my courses is to help the student learn to solve rather complex problems by using relatively small computing power, such as personal computers.

The acquisition of computers poses serious problems in the poorest of the developing countries. Does this discourage teachers, policy formulators and decision-makers from thinking 'system theory'?

I hope not. Thinking in a systemic way should come first; the computers can follow. As small personal computers become cheaper, they become affordable even to small developing countries. Joint research teams, representing two or more countries sharing the same kinds of problems, may be a profitable solution to the problem of procuring computers.

This is networking, itself an application of system theory, but does it work?

Yes, and our ability to stabilize electric power systems is a good example of this. In Yugoslavia, as our power network becomes more closely connected with those of other European countries, problems of system instability can be expected to arise.

Instability can follow if, first, the largest of several power plants shuts down. Secondly, if consumption begins to outrace production. Thirdly, if an incident occurs in the 'transition network'. The problem is how to control the entire system and preserve its stability while dealing with a perturbation and the same applies to control of nuclear power-plants.

In the case of the accident at Three Mile Island in the United States, there was no shortage of warning information; there was simply too much conservatism in applying new technologies where these could maximize the system's stability by exploiting the information available.

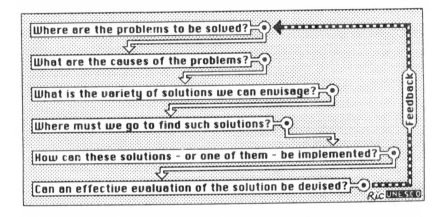

In other words, all feedback loops must be closed to ensure a fail-safe system?

Indeed, and this is just what we built into our model for Yugoslavia's first nuclear power plant. We insisted on robustness in the system. Let me explain robustness.

In science, we are made aware of the problem of sensitivity when we try to preserve a given property in a system having infinitesimally small variations. For example, the maintenance of stable flight by an aircraft demands instant sensitivity of the entire system to minutely variable factors. Robustness as a problem is in some ways akin to that of sentsitivity because, in the presence of large-scale perturbation, we are trying to preserve a system's performances, including its stability. This is one aspect; another aspect of robustness is the real difference existing between mathematical models and the system they depict. We need a systemic approach to define the difference between a model and its real object, thus defining the cause or consequence of differences established concerning the model and its real system.

Are mathematical models good approximations of how a system works?

We have been using models in science for a long time. When we began to apply modeling to economics and the other social sciences, perhaps we were expecting too much. There are possibilities, but there are limitations, and writing equations is not all there is to model-making. One person alone cannot make a valid model; the work requires an interdisciplinary team. Failed models are inevitably attributable to lack of working together in teams.

You are currently concerned with some of the interdisciplinary problems confronting the countries of the Mediterranean basin. . .

There are multidisciplinary considerations related to the dynamics of the Mediterranean Sea (itself a large-scale system) and the problems along the coastal regions, most of which are man-made. New approaches and new methods are devised in one region, but still unknown in another. This is where international co-operation in research and development pays so well: helping people to think in a new way, to view a problem in a different light. This is how we solve problems effectively.

What kinds of systems would you like to be associated with in future studies?

Economic systems, especially in terms of national planning. I wish to consider Yugoslavia as a large-scale system, with all the interconnections this implies, and to apply some of the latest ideas of large-scale system theory to economic planning at the national level. I realize that making such simulations requires people, time and much experience—the last of which (as I have already said) can often be learned from others' efforts. I do not wish to imply that making a good model, full of equations and mathematical symbols, is the only way to solve real-life problems. I am not yet sure what the next step should be, but modeling on a global basis should be very helpful in our efforts to work out things on the national scale.

Simplicity derived from complexity?

Yes, but then the next step may involve movement towards some kind of non-linear system. As a problem becomes more complex, we tend to develop new mathematical tools: field theory, or dynamic interactions. I am not sure that making the mathematics more complex is the right way of tackling a problem.

Are social scientists willing to be involved with mathematicians, engineers and natural scientists working on complex matters?

We cannot apply large-scale theory to even a simple system problem without considering sociological factors. We do, indeed, need specialists from the human sciences, but unfortunately there remains a gap between us when we seek a common language to describe such problems. This is in itself a system problem to be worked out in the future.   ■

Note: This interview appeared originally in the English, French and Spanish editions of *Unesco Features*. It is reproduced here, with minor modifications, with the kind permission of the editors of *Unesco Features*.

See also:

SILVERMAN, B.G., Toward An Integrated Cognitive Model of the Inventor/ Engineer, *R&D Management*, April 1985.

Language problems impede international understanding, but in recent years computer developments have provided us with a possibility of averting some of these. Studies of computer languages have not only begun to break down national language barriers, they are leading to a deeper understanding of the way people communicate by natural languages, and of the way we think. Thus technological innovation may one day lead to increased international accord.

Chapter 18

The critical role of language in the man-machine relationship

Ouyang Wendao

Dr. Ouyang was originally an economist and administrator who went first into engineering and then into scientific research, most recently in artificial intelligence and language processing at the Institute of Automation of the Academia Sinica, Beijing, China. The author is a council member of both the Esperanto Association of China and the Chinese Information Processing Society. His address is: Institute of Management Sciences, Academia Sinica, Post Office Box 3353, Beijing (People's Republic of China).

It would not be an exaggeration to say that linguistic issues have joined the ranks of the world's most important and challenging problems. Currently, interest in such problems has increased with the introduction of the electronic computer into language specialists' studies. The probable impact on society is impossible to predict, but it will depend to a large extent on how creatively innovations are introduced in this field.

Language: a medium of communication

No matter how well developed modern communications media are, the most fundamental of them is language. Today 2,500 to 3,000 different languages are spoken by the peoples of the world. The eleven most widely spoken rank in the following order: Chinese, English, Hindustani, Russian, Spanish, French, Portuguese, Arabic, Indonesian, Japanese and German.

Each of these languages bears national, historical and cultural characteristics, and is used within its country of origin without hindrance. However, these languages constitute a communication barrier to people who do not belong to the same language group.

Take Chinese as an example. On the Chinese mainland alone, 1,100 million people speak and write Chinese: the Chinese language is the melting pot of Chinese history and culture. But although Chinese is one of the six working languages of the United Nations few foreigners, except for a few experts and scholars, can master it. This inevitably leads to little being known about China beyond the country's borders.

An analogous situation arises with Chinese people studying foreign languages. Those currently learning them do so with great enthusiasm. Not only is English studied but French, German, Japanese and many other languages. Yet while those who study these languages number in the tens of millions, only a few will be able really to master and use them freely. Thus, hundreds of millions of Chinese will know little about countries other than their own.

Human contacts have increased rapidly with the development of modern communications and transportation. But language barriers still prevent people from understanding each other. Such a situation works against world peace and social progress.

In the past, Bible stories were used to describe language barriers as 'natural' problems (e.g., the 'Tower of Babel'), and to make such problems seem almost insoluble. But the march of civilisation has shown that natural problems can indeed be solved, and that nature can be modified to permit the development of mankind.

Thus today many people with insight are calling for solutions to linguistic problems, for example through an international language that would be used by every person in addition to his own national one. Such a language should permit a richness of expression, and would help eliminate prejudice and discrimination based on language. It should be easier to study and master than national languages: Esperanto is an example which provides a practical solution to the problem.

Machines that can communicate with man

With the widespread use of electronic computers, a society that is highly dependent on information distribution has emerged. There is little doubt that the development of artificial intelligence exemplified by these machines will have a profound effect on society. Even beyond the question of a common language for all mankind, this development raises the possibility of a common language of communication between machines.

Machines, it is now clear, will become not only man's mechanical slaves but also his close associates through which humans can communicate with each other. The newly developing science of artificial intelligence is now being dedicated to studies of natural language processing—understanding man's language through computers—and has already obtained some preliminary results. In this way, fifth-generation computers may bring to realisation a dialogue between mankind and machine through natural language.

Such work is already going on in China: the Information Processing Society of China (CIPSC), which was founded in June 1980, organizes and promotes such research throughout the nation. In recent years, we have built computers that can input, output, store and manipulate several thousand Chinese ideographic characters as well as alphabetic letters. We have also designed Chinese language information models that can help computers identify Chinese words and phrases from a series of Chinese characters used to compose legal sentences, carry out syntactic analysis and formulate semantic representation. In short, in our laboratory we are managing to build computer systems that might be able to help foreigners to comprehend the Chinese language, which has seemed so difficult for them in the past. On the basis of this arduous work, we might in the future be able to develop a machine that could help interpret various other languages. This of course would be an ambitious project that would necessitate scientific collaboration both within China and internationally.

A new proposal has been presented to the CIPSC to build a knowledge-based Chinese information system so as to seize upon the opportunity presented by the development (in the United States and Japan) of the fifth generation of computer technology. It has also been proposed that a new center for Chinese-language processing be set up soon. Scientists and technologists from several institutes are working together along these lines.

Language: a cultural mediator

The language that appeared first in computers was the language of the machine itself—'machine language.' This was followed by high-level program language, an artificial language common to those who program computers. While not a language of any single nation, program language necessarily is based on the languages of those nations that use it. This raises the question: In the computer world, will a commonly used international language be necessary in the future?

Some studying machine translation systems have already proposed basing such a system on a kind of intermediary language which would deal with many national languages. Such a proposal seems reasonable. Others suggest basing this intermediary language on an internationally accepted human language such as Esperanto.

Whatever happens, we can imagine that some day it will not be necessary for those who speak, read and write their own languages to study many foreign languages. For they will be able not only to communicate with nationals of other countries in person through one international language, but will also be able to promote mutual understanding by means of machines. With the help of an intermediary language, they will be able to read freely and easily material appearing in many national languages.

When that happens, the Tower of Babel will soar aloft, built not of bricks and stones but of human wisdom, friendship and patience. Such a tower would not be unnatural, but rather the highest expression of wisdom yet evolved.

Language: a medium of thought

The demand for natural language processing by machine promotes not just computer science but also linguistics. Neuropsychologists have to study the subject to see how computers can simulate the cerebrum's language-based information-processing functions. The relationship between language and thinking has long been a philosophical and

Fig. 1. The complexity of precise expression in certain languages is typified here in contemporary Chinese. In order to convey the abstraction 'crisis,' two other substantives must be compounded: 'danger' and 'opportunity'. A similar situation exists in the modern Japanese and Korean languages and, to a certain extent, in contemporary Arabic. Note the positive or constructive connotation of 'opportunity', whereas in English and related European languages the term 'crisis' has a negative intimation. (From a seminal idea appearing in L'Intelligence transformatrice, an unsigned article in La Revolution de l'intelligence (theme), prepared by the editors of Sciences & Techniques, *special issue, March 1985.)*

scientific concern of mankind. It is generally considered that language is not only the communicator of thought but also its very foundation.

In order to process language, the computer requires that language be analysed and described with great accuracy by man, and that its structures and rules be revealed in unusual depth. But computers also make it necessary for many to reflect deeply about the brain's methods of processing language. Such studies will not only deepen our understanding of language, but also of thought processes. Some scientists even consider that a scientific system of thinking is beginning to take shape. It is a conjecture not without foundation.

Words are both a kind of model and the symbols of ideas, and language describes the relationship between one idea and another. Thought comes about through the mental manipulation of these symbols. In M.N. Quillian's opinion, ideas exist in man's memory, and people communicate with each other and understand what others say on the basis of these stored memories. The significance of ideas on the understanding of language has been proved by S. Rosenberg in his psychological experiments. When one comprehends a sentence, it is the idea and the semantics, and not the grammar, that play the major part. The conceptual dependency model of ideas proposed by R.C. Shank is a model of a comprehensive computer language built on the psychological basis.

Language comprehension or language expression

Our own studies demonstrate that the characteristics of national languages are primarily reflected in the surface and middle layers of language (that is, on the level of words, phrases and syntax). Once the outer layer is peeled off, the deeper layers, containing the ideas, are exposed. The comprehension of language is thus a psychological process proceeding from the outside to the inside, but the expression of language is precisely the opposite: a process proceeding from the inside to the outside. These two processes often occur alternately in communication by language, which illustrates how language and thought are interwoven.

Doubtless, studies on the processing of natural language will enhance man's ability by means of computers to master the medium of language, thus inevitably increasing the efficiency of language as a medium of thought.

Man's mastery of language and consequent improved ability to think would have a profound effect on society. More significantly, with the common use of an international language, people everywhere could share more fully with other nations their scientific, technical and cultural accomplishments. In such a situation, highly favourable conditions would be created for a reasonable worldwide utilization of mankind's material and spiritual resources, and for the solution of a wide range of international problems.   ■

See also:

PORTNOFF, A.-Y., Les industries de la langue existent (There is A Language Industry), *Sciences & Techniques*, No. 20, November 1985 (new series).

What part does language, man's most effective tool, play in the process of learning to be creative? An American scientist of Japanese ancestry, who has spent many years studying the relationships between culture, language and scientific creativity, attempts to answer the question.

Chapter 19

Pattern recognition
and its relation to creativity

Roy Katsumi Uenishi

Dr. Uenishi obtained his advanced degree in chemistry from Pennsylvania State University. He has done research in a variety of fields, including rocket-fuel oxidizers and laser-initiated chemical reactions, and is currently concerned with creativity in science enhanced by artificial intelligence. The author is a self-employed inventor, translator of Japanese patents, and registered patent practitioner who can be reached at 1255 19th Street, No. 402, Denver CO 80202 (United States).

Rapid extraction of meaning

Japan, not surprisingly, has already distinguished itself in biochemistry, microbiology and some medical sciences where ready recognition of interrelationships between complex patterns and groups (and synthesis) is frequent—the very attributes that distinguish ideographic languages. Training in these sciences involves considerable pattern recognition (implying memorization). They are provisionally more amenable to inductive reasoning, as in the empirical arts and sciences, and emphasize perception (i.e., quick, acute and intuitive cognition resulting from observations), though deductive reasoning plays a role later to check hypotheses.

The problems of such sciences are more complex than those of exact sciences dealing with disorganized matter. Intuitive cognition, with its holistic grasp, can effectively handle organized—though not identical— repetitive entities: myriad living cell structures with dynamic compositions, extensive arrays of biochemical structural formulae, any complex patterns and groups that signify thought, as do ideographs. Not infrequently, intuition even supersedes analysis, which ordinarily precedes complex synthesis.

Japanese *kanji*, borrowed from Chinese logographs, were in general fashioned to resemble the objects they described—the observer feeling more a part of, than apart from, phenomenal nature. Because the meaning of logographs does not change with time, though the sounds can, thought symbols were communicable from Chinese to Japanese. However, because of the phonic characters (*kana*), the reverse is not necessarily true. One does not construe a Japanese sentence sequentially as in alphabetic sentences (each word standing independently, distinctly, and in clear relation to other words): it involves more pattern recognition and gestalt. Reading in *kanji* probably involves greater overlap of graphic, phonemic, and semantic aspects of word recognition than in alphabetic languages or *kana*.

Thus, in spite of its graphical complexity, the meaning of *kanji* can be extracted at a glance. Also, readers of Chinese logographs have been found to possess better visual recall than readers of English. Studies using meaningful complex visual patterns and groups of such patterns (e.g., histological slides) to test the relative visual recall of Chinese, Japanese, Korean, and English readers would be informative. The 'one-step synthesis method' of reading Japanese, which emphasizes synthesis and rejects analysis, is also of considerable interest for those designing fast translating machines.*

Calligraphy, script, and euphemism

Alphabetic words are phonetically based and their meanings change with time. This indicates a significant difference from logographs and one that bears on one's world view, showing on the part of alphabet-language speakers a greater awareness of the rate of change and a need to understand nature rationally. Apparently all written languages originated from their respective spoken tongues—all are models used by humans to simplify the reality they represent, all using pattern recognition, analysis and synthesis. It is no less apparent that the emphases are different and consequently the thought forms are not the same. Therefore the manner in which the cerebral hemispheres are used must be correspondingly different, thus influencing the manner in which genius (inventiveness, creativity and originality) is brought out.

Both German and English grammar consist of morphology (structure) and syntax (word order). However, German is more structured and English more flexible, in that English has more allowable ways of expressing thought symbols in easily understood verbal forms, allowing greater freedom of judgment by the individual. Significantly, the laws of the exact sciences tend to be rigid, requiring logical grammatical structure and word order, while documentation requires critical judgment by the writer.

Japanese thought form, i.e., expression of judgement and deduction, is regulated by the language form, i.e., grammar and syntax, and has undergone revolutionary changes due to the upheaval of the Second World War followed by unprecedented exposure to Western languages. The Japanese language is a *kanji-kana* hybrid and does not exhibit the grammatical excellence of Western languages, but this hybrid, plus the invariant pronunciation of *kana*, simplifies the design of translating machines and facilitates access to artificial intelligence. Contextual implications (semantic and syntactic) and tacit understandings are

---

* This aptitude occurs frequently among Japanese and is probably a consequence of greater overlap in Japanese word recognition than in other languages (i.e., graphic, phonemic and semantic recognition) as well as in gestalt recognition.

critical in construing Japanese sentences but referents are not always explicitly clear. Ambiguities are not infrequent.

A language heavily dependent on memorization of complex patterns and calligraphy, euphemism, and the expediencies of romanized script of limited general utility because of imperfect transliteration and numerous homophones, probably tends to structure a certain type of educational system. Besides language aptitude, it also no doubt promotes certain traits among science and technology graduates: adaptability, determination, perseverance, and specific goal-orientation are prerequisites.

The confining tendencies of Japanese may actually provide advantageous conditioning in the highly competitive and specialized age of today. The imperfect yet workable adaptation of Chinese logographs together with the *kana* (themselves based on the innovation of the logographs) are relevant to Japan's startling rate of adopting/'adepting' Western technologies.* What relevance does the Korean *hangul* syllabary have? This is a question that remains to be answered.

Logographs, bookishness and conformity

It is popularly acknowledged that a greater percentage of Japanese pre-college students are studying English and calculus today than at any time in the past. I have yet to meet a Japanese scientist without some working knowledge of English. Many are adept. Significantly, Julian Huxley ventured that language acculturation may well displace biological processes of evolution. But Japanese science is not necessarily just an imitation of Western ethnocentric science. Japanese ethnocentric science may be evolving.

T. Tsunoda has scrutinized the approaching transformation of creativity and originality in Japanese sciences.[1] The predisposition of Japanese students towards bookishness—extensive library research and arriving at an authoritative consensus in contrast to finding answers by trying a radically different approach through independent thinking— may be due partly to excessive use of the left (linguistic) cerebral hemisphere and not enough use of the imaginative and intuitive functions of the right. Tsunoda's studies, for example, seem to show that children under 10 years of age and acculturated through Japanese will use the left hemisphere more than Westerners or other Orientals. Sounds of nature (wind, waves, animal and insect sounds, etc.) are heard by Japanese researchers of the social structure of monkey colonies as vowels,

---

* Adopting-adapting-'adepting' is the usual sequence of processes in the importation of a foreign culture, *viz*, western technologies by the East.

and are processed in the left hemisphere. A deeper insight is believed to result.

Day-dreaming, relaxed contemplative (*zazen*) or hypnagogic (drowsy) states of consciousness trigger access to both the left and right hemispheres of divergent thinking, i.e., creative/imaginative thinking. August Kekulé and Norbert Wiener frequently used concept models derived through the hypnagogic state. Contrarily, coffee, tobacco, citrus juices and certain chemicals (e.g., ammonia) are conducive to linear or sequential thinking. This results in access to the relative predominance of left hemispheric activity during convergent (i.e., rational) thinking. Convergent thinking thus clearly complements divergent thinking in the sciences; the former provides hard facts from experimental work or rigorous thought to back up the ideas of the latter.

Finally, the conception of truly original ideas entails non-linear thinking and holistic consciousness of the individual, necessarily diverging from established patterns in order to create different inter-relationships between seemingly contradictory patterns and groups.

Japan's social conformity is at least partly perpetuated by its language and is constructively expressed in the mobilization of skilled labour in such areas as microelectronics. Written Japanese conditions the mind towards group cohesiveness by obscuring the individual. Often, *kanji* and *kana* are indiscernible in the text, but in reading such words combining *kanji* and *kana*, the meaning 'pops out'. Social obligations, duties, and respect for parents, elders and authority are of Confucian origin, hierarchical, and reinforced by honorific protocol in speech and writing.

Mastery of Japanese calligraphy enhances awareness of Japanese artistic skills, precise and intricate use of the hands, personal cleanliness and neatness, respect for details, self-discipline, perseverance for perfection, imitativeness, creative synthesis and flow (notably *sumie* brush painting and *zazen*), proper form and procedure (e.g., flower arrangement, tea ceremony, all rituals, software programming), group dependence (i.e., adaptability) and group decision-making (the whole must be greater than the sum of the parts). All these are positive traits applicable to the regimented assembly-line mass-production and quality control essential in microelectronics. Though contemporary college graduates are by no means calligraphers, they have been exposed to the stringent requirements of calligraphy during pre-college schooling.

A look ahead

The revolution in biology should lead to nothing less than the second stage of human evolution. The first stage has taken a million years or so,

and has been dictated by chance. The second stage should take less than
a century and will be dictated by necessity.[2]

With the rapid extraction of meaning inherent in the reading of
ideographic languages, together with (a) the connection between pattern
recognition and the creative act and (b) the linear thinking of alphabetic
languages inherently capable of rigorous abstract thought, the artificial
languages of future computers should amplify man's creative acts by
synthesizing the two ways of thinking, alphabetic and ideographic, and
thus accelerating the reality of the second stage.	■

Notes

1. T. Tsunoda, *Nihonjin no nō* (The Japanese Brain), Tokyo, Daishukan, 1978.
2. C. Renmore, *Silicon Chips and You*, New York, Beaufort Books, 1980.

To delve more deeply

BODEN, M. *Piaget*, Brighton, Harvester Press, 1979.
CHENEY, M. *Tesla: Man Out of Time*, Englewood Cliffs, Prentice Hall, 1981.
GIL, F. Scientific Creativity in Today's Society (working document, Programme
    on Research and Human Needs), Paris, Unesco, 1982.
PAIS, A. *Subtle is the Lord: The Science and Life of Albert Einstein*, New
    York, Oxford University Press, 1982.
TORRANCE, E. *Guiding Creative Talent*, New York, Prentice Hall, 1962.
WATSON, J. *The Double Helix*, New York, Atheneum, 1968.
WIENER, N. *Ex-prodigy: My Childhood and Youth*, New York, Simon &
    Schuster, 1953.

In this chapter a concise review is made of the origins of computers and computer programs and what they do, what computers have substituted for in history, and the computer-human relationship and its pitfalls. The narrative focuses on the interaction of children and the computer. An assay is made of the most rational future uses of both the computer and artificial intelligence.

Chapter 20

The computer—tool, helper, friend—and the future of artificial intelligence

Jacques G. Richardson

The author prepared the original version of this analysis of one of man's most creative implements for an international conference and exhibition, Children in An Information Age: Tomorrow's Problems Today, held in Varna (Bulgaria) in May 1985. The title of this meeting helps to explain the style of presentation of the material.

. . . people's participation in decision-making is the *sine qua non* of development. It is also clear that children play a key role in that process. They serve as the motivating force for improving family and community well-being. With a brighter future for their children as an incentive, parents strive to improve the environment for growth and take new technologies and ideas in their stride.

— World Health Organisation [1]

What is a computer (and its program)?

This device is very much like a carpenter's hammer or an electrician's screwdriver. It assists, at your fingertips as well as in the workings of your mind, how you count; how you store numbers, words, and ideas; and then how you combine any of these to tell a story, draw a picture, or solve a problem. It helps you to predict—to say in advance—how combining words, counting and using ideas will be useful to you, your family, your friends and your whole community.

When he was only 19 years old, a little more than three centuries ago, the French thinker Blaise Pascal devised an 'arithmetic machine'. This was a simple prototype capable of adding numbers, barely an improvement on the Chinese digital abacus already in use several thousands of years. A mathematician and physician, Pascal had little practical purpose in mind when he conceived his simple calculator other than to facilitate computation.

A little more than a hundred years later a countryman of Pascal's, Joseph-Marie Jacquard, sought a simplification of the complex process of forming design while weaving silk. Jacquard represented all the steps involved in weaving on cards carrying different combinations of holes. As the joined cards passed through a specially designed loom, their holes held back or allowed to pass various combinations of rods bearing thread of different colours. What Jacquard achieved, besides saving tedium and time, was to define a mechanical process by means of the information contained in its constituent movements: this is the basic process inherent in the computer.

Not long afterwards, in Great Britain, Charles Babbage, another mathematician, contrived an 'analytical machine' capable of weaving patterns of algebraic expressions much as Jacquard's machine could weave artistic patterns. Babbage was assisted by another mathematician, Ada Byron (Countess of Lovelace), who elaborated the first instructions or *program* for Babbage's *computer*.[2] In the 1890s, the perforated card was used the first time to compile statistics on a voluminous scale: the national census of the United States. Carded information became a popular time-saving feature in much office and laboratory technology until the end of the Second World War. This technology included logarithmic tables and the slide-rule.[3]

The modern computer made its appearance at the end of this conflict—partly as the result of military needs for complicated calculation having to do with aerial flight, sighting targets and controlling nuclear energy, partly as a result of the invention of the transistor in 1947. The new instruments were the first calculators to process information transported by electronic flow, data treatment which has been speeded up many times since as new technology has allowed repeated miniaturization of the equipment used. The design or writing of programs has also become far more complex than it was a mere 30 years ago, having given birth during this time to the creation of special languages enabling man to communicate with the computer, and vice-versa.

The computer of the 1980s is capable of handling virtually all kinds of mathematical procedures, memorizing them as well as millions of stored words (there are entire libraries assembled within a single computer system) and feeding back such information when called for.

Today, using a computer, it is possible to draw a picture, write a poem or a book, calculate the precise location of Halley's comet one month or one year from now as well as the universe's rate of expansion, to design a new ball-point pen or an automobile or a skyscraper, to conceive a new plastic substance or a genetic structure—and even to make reasonably valid predictions of the weather and farm production, intricate projections of currency flow, and modelling of natural cataclysms to come.

What did we do before there were computers?

We memorized: simply, by using a stick to write in the dust, a nail to write in wax, chalk, a pencil and then a typewriter or printing press. To count, we often used much paper and spent much time. A famous Danish astronomer, Tycho Brahe, filled 9,000 sheets of paper to estimate (for the first time) how the sun's planets move around it; after many years of work, he found that his results were wrong. Writers and artists took similar pains, often having to start again from the beginning. The computer helps to shorten time, make difficult tasks easier.

Evidently, the computer and what it can do for us are very much a part of the modern age. If it has proven so practical, how did mankind get along so long without this useful instrument?

Counting. Man first learned to enumerate objects in pairs, by threes, and so on. He learned to store simple sums on his fingers and in the mind, eventually to combine operations so as to add and subtract, multiply and divide. The Indians developed the concept of zero, the Babylonians and Egyptians contrived empirical geometry and the Greeks 'theorized' it. The Babylonians, Chinese and Arabs devised algebra, later refined by Descartes and Boole. Trigonometry had similar early origins, later refined by Newton who, concurrently with Leibniz, invented calculus. The last is the Latin word for pebble, suggesting the extent to which our ancestors depended on small natural objects for the modelling of their computing. Our century's contribution to mathematics has been the development of operational analysis, a combination of mathematics and engineering orderliness.

Writing. A craftsmanship originating probably in the sketching in dust or sand, with the fingers, of nearby nature and its wonders, this ability was expanded to the paintings 25,000 years old found in caves in Spain and France. Egyptians and Romans, prolific writers, recorded on stone, papyrus and wax tablets, whereas parchment probably originated in Asia Minor. Wood-fibre paper came later, derived originally from Chinese bamboo pulp; it was brought by Islam from Turkestan through North Africa and Spain to Europe and the rest of the world. Writing instruments of all kinds followed over the centuries, culminating in our epoch's computer.

Literacy. This includes numeracy and is, today, widespread chiefly in the industrialized countries. Illiteracy affects a society's ability to learn (and, therefore, to teach) other than empirical knowledge and folklore. Limited knowledge affects, in turn, the development process as a whole: economic change, technical advancement, social evolution. Where the development process is slow, the human memory must work overtime, whereas in industrialized societies the memory's functions have been taken over in part by the written word and mechanized counting. The informatisation of the development process is thus a key, in itself, to development, one often regulated—as it were—by demographic patterns: Whereas literacy may rise in absolute numbers, a quickly growing population often offsets the progress made to overcome analphabetism and low rates of schooling and vocational training.

The computer and the informatisation of society are by no means a panacea in the case of these societal problems, but concerted efforts to raise the rates of numeracy and literacy are inevitably contributing factors in the amelioration of national development.

Is the computer a friend?

Of course, it is. We see how it helps us to do our work faster, more accurately, less tiringly. This tool gives us pleasure in games and puzzles, too. But the computer's little screen can also invade the time we should use for other things: games on the playground, sports and having fun with our friends, reading, playing a musical instrument, learning how to act or paint or build things.

'Rather than harming children's education, the new (electronic toys and games) stimulate their interest', according to Huang Shouchun of the Shanghai Children's Palace.[4] The specialist added that in 1983 an eight year-old schoolchild won a prize for inventing a simple machine to prevent short-sightedness in children. 'Placed on a school desk, the device sounds an alarm and flashes when a pupil gets too close to a book' while reading. 'The young inventor said he got the idea from an electronic toy dog that barks when people approach it'.

Soviet schools distributed in 1985 a total of 1,131 Soviet-made *Agat* personal computers to ninth-grade students. In 1987, computer instruction commenced at the fifth grade, as part of a long-term plan to bring computer literacy to the young.[5] In the United States, some university-level institutions require that each student have access to his or her own computer, with already a half-million computers in use in high schools and primary schools, although much teaching remains of the paper-and-pencil sort—according to the chairman of the United States' National Commission on Industrial Innovation. 'Computers uniquely change the potential for equipping today's citizens for the unprecedented tasks of the future', this specialist remarked. 'How successfully we respond will depend on how much we invest in people and how wisely we employ the learning tools of the new technologies'.[6]

The computer and its putative effect on children and youths, is interactive, fast, 'friendly'. It may form or reshape the life of a youngster in unexpected ways, as depicted in very popular films such as *War Games*. The computer, according to MIT's Professor Sherry Turkle, 'can influence people's conceptions of themselves, their jobs, their relationships with other people, and their ways of thinking about social processes'.[7] But she adds that 'the satisfactions that the computer offers are essentially private'. In societies where a sense of collective welfare is a paramount concern, egoistic pleasures will have to be reconciled with common needs—*pro bono publico*. This is a socio-technical challenge of considerable dimension.

How can we make the best use of the computer in the future?

As this attractive, useful and co-operative machine improves while the child grows, so must his attitude about how to use a data processor be guided by his parents, his teachers and his playmates. The youngster needs to learn that he will always have the power to guide the computer, to use it selectively—to dominate the instrument at all times.

With the advent of supercomputers operating at extremely high speeds and the appearance in the late 1980s of artificial-intelligence machines,[8] new computing horizons are opening. These include health-problem diagnosis and other 'expert systems', direction of intelligent robots, transcription of dictation and the translation of languages. The computer can be expected to become ever more co-operative and 'user-friendly'. It will even assume a predominant role in the user's daily life.

Inherent limitations to artificial intelligence.

. . . while it's relatively easy to program computers with skills that humans have to learn, it's very difficult to program them with the skills we acquire unconsciously. This shouldn't be too surprising, for no one has access to the unconscious mind. It functions automatically, at a level we are only beginning to understand. The physical processes that enable us to see, to remember, to carry on a conversation, to function in the everyday world—all these remain a mystery.

The other difficulty is that in order for a machine to function in an unrestricted environment—that is, the real world—it needs to have a complex model of what the world is like. In order to play chess, a computer doesn't need to know about anything but chess. But to carry on an intelligent conversation . . . about, say, fishing, it needs to know about fish, people, sport, water, weather, worms, and getting up at five in the morning, not to mention the difference between a riverbank and a savings bank and what any given pronoun is supposed to refer to. In other words, it needs to have a vast array of knowledge about the world, and it needs to have it organised in such a way that it can be accessed instantly on demand.

Frank Rose, *Into the Heart of the Mind*
New York, Harper & Row (1984), p. 50

Whether or not the electronic calculating machine will become the computer Hal of the film *2001 Space Odyssey* remains to be seen. Meanwhile, however, man remains very much master of the machine, and it is incumbent on instructors and other adults to help remind the learner on the computer of the device's inherent limitations and of his or her final control over the computer and its programme.

Equally important are considerations related to ethics in the young person's application of his knowledge of computers, their programming and use. This ethics includes protection of data in terms of privacy (whether institutional or individual), 'computer crime'—clearly anti-social acts in the use of the computer that are or will be punishable by law, and 'computer politics'. The last is a phenomenon known so far only in a few industrialized nations. Finally, ethics also embraces the question of cross-border data flows, a phenomenon with international cultural and political implications.

Seymour Papert, a disciple of Jean Piaget and creator of the LOGO computer language and for a while a director of France's Centre Mondial d'Informatique, has expressed tersely his own moral attitude in the computer-youth relationship: '. . . I believe that certain uses of very powerful computational technology and computational ideas can provide children with new possibilities for learning, thinking, and growing emotionally as well as cognitively' in 'a future where a computer will be a significant part of every child's life'. [9]   ■

Notes

1. *WHO Features*, No. 90, November 1984, p. 5.
2. See W.J. Eckert, *Punched Card Methods in Scientific Computation* (1940) and D. Hartree, *Calculating Instruments and Machines* (1949), Vols. V and VI of the Charles Babbage Institute Reprint Series, Cambridge, MA, MIT Press, 1984; and J. Weber, *Tomorrow's World: Computers, The Next Generation*, New York, Arco Publishing, 1984 (from the BBC Television series).
3. See La saga de l'informatique (theme), *Sciences & Avenir*, special issue No. 49, 1985, for a good popular presentation of the evolution of automated data processing.
4. Reported by the Xinhua Agency in *China Daily*, 19 September 1984.
5. W.J. Eaton (Los Angeles Times Service), *International Herald Tribune*, 21 September 1984, citing a Soviet teachers' journal.
6. Former California governor Edmund G. Brown, Jr., in the *International Herald Tribune*, 26 February 1985.
7. S. Turkle, *The Second Self*, New York, Simon and Schuster, 1984.
8. A good survey is Gene Gregory, The Rush Is On to Develop Ultra-High Speed Machines: Japan on the Fast Track, *Far Eastern Economic Survey*, 11 January 1985.

9. S. Papert, *Mindstorms: Children, Computers and Powerful Ideas*, New York, Basic Books, 1980; Brighton, Harvester Press, 1980.

To delve more deeply

Anon., *Artificial Intelligence: A New Tool for Industry and Business,* Ft. Lee, NJ, Technical Insights, 1985.

Artificial Intelligence (feature section), *Science 85*, March 1985. T. Forester (ed.), *The Information Technology Revolution*, Cambridge, Basil Blackwell, 1985. A 'complete guide' to the subject.

James Burge (producer and director), *Thinking*, a 52-minute television production in the Horizon series, BBC-TV, first broadcast in February 1988. An excellent popular treatment of our present knowledge of the thought processes, or 'finding the program of the mind.'

L. Gonick, *The Cartoon Guide to Computer Science*, New York, Barnes & Noble, 1980.

C. Norman, *The God That Limps: Science and Technology in the Eighties*, New New York, W.W. Norton, 1981. A review by a science journalist of global questions arising from the revolution in information technology.

La Révolution de l'Intelligence (theme), *Sciences & Techniques* (special issue), March 1985.

L'intelligence artificielle (theme of special issue), *La Recherche*, October 1985.

E. Rich, *Artificial Intelligence*, New York, McGraw-Hill, 1983.

R. Trappl, *Impacts of Artificial Intelligence: Scientific, Technological, Military, Economic, Societal, Cultural, Political*. Amsterdam, Elsevier North Holland, 1986. Dr. Trappl is at the Austrian Research Institute for Artificial Intelligence, in Vienna.

A Chinese nuclear engineer and perennial investigator in comparative research and development examines the prospects for innovation in China in the coming years. He analyzes traditional cultural concepts, the country's production system, as well as the imperative to develop new institutions in order to aid and abet the nation's innovative output based on scientific technology.

Chapter 21

New scientific technology and its influence on decision making and management in China

Wu Jisong

Mr. Wu (born 1944) was educated at Beijing's Tsinhua University, his country's leading technological institute. He did research on nuclear fusion for the better part of two years at Fontenay aux Roses (France). As a senior fellow of the International Federation of Institutes for Advanced Study, he travelled extensively in Europe, North America and Japan in order to study planning and management in research and development. The author returned to China in 1986, where he is deputy chief of the Division of Planning and International Co-operation, Chinese Academy of Sciences, 52 San Li He Road, Beijing; telephone 863062.

Background

New scientific and technological processes and products always make a strong impact on social development in any country, during any historical period. They stimulate further creativity and invention, which can be viewed as main driving forces for the further development of society.

During the European Renaissance, print technology was introduced to Europe from China. There was also an improved, medium-movable type in Korea, with the result that a new facility in printing created astonishment in intellectual circles. People began to believe that writing would no longer convey personal sentiments because uniformized, printed letters would replace individual manuscript. This is a good example of the impact of technological process on the history of socio-cultural evolution.

Today, the impingement of new scientific technology on development is an extremely important issue in the so-called Third World. The discussion in this chapter concentrates on the consequences of new scientific and technological processes for decision-making and management in the sphere of research and development (R&D) in China. Although these consequences are manifold, we limit our discussion here to three areas: the production system, traditional and new concepts, and the development of institutions.

Impact on the production system

Further scientific and technological processes and their products will continue to have a direct effect on the production system in developing countries.

First, a traditional advantage in Third World nations has been the combination of a cheap labour force and rich resources in raw materials and energy. This advantage becomes considerably limited in an era of new technology based on science. Electronics, biotechnology and industries using new materials do not depend as much as previous industrial activity on labour and primary resources. (The iron and steel, automotive and petrochemical industries rely heavily on these.) A widespread use of robots to replace labour-intensive technologies will probably prevail in the near future, even in the traditional clothing sector. There is a keenly felt need, therefore, for reform of the conventional industries in developing countries and for the encouragement of developing new capacities in science and technology.

Secondly, because of these reforms and the advancement of science and technology, the employment problem will become seriously aggravated. There is, on the one hand, a critical shortage of skilled workers and, on the other hand, an excessive supply of unskilled hands. The tasks ahead are therefore not simply to upgrade the basis of normal education and training, but to popularize as well the critical need for vocational instruction throughout the countries of the Third World.

Thirdly, there remains a contradiction between new science and technology and a country's backward infrastructure (basic facilities). Information technology, for example, requires certain conditions in the electronics industry, the electric power-supply system, and the telecommunications network. The adaptation (or the outright adoption) of any kind of information technologies should be based on such considerations. Thus, the selection of priorities in the import of new technologies—or their development at home—should be based on a combination of system analysis and feasibility studies.

Impact of concepts

One of the kinds of impacts that scientific discovery and new technical processes have on Third World development is not always sufficiently considered. This is the effect that they have on concepts, or received ideas and the conventional wisdom. Because of space limitations, I shall mention only a few examples of these.

The value of information

Looking over the history of R&D, one conclusion that can be made is that this history depends to a large extent on the means available for the exchange of information. I have illustrated this sort of relationship in Figure 1.

There are impacts of information transfer on all kinds of human activity because all interactive processes depend on (a) communication in general and (b) the exchange, in particular, of what we now call scientific and technological information. Each improvement in technology has progressively removed the constraints on communication between individuals and social groups, moreover, and (inversely, as it were) this progress has provided new conditions for the R&D process.

Today, in all developing countries, the need is keenly felt for a large and accurate volume of data, a high speed of retrieval, and a wide-ranging sharing of information as well as broad support to information centers. Satisfaction of this compound need is essential in order to answer queries related to all kinds of R&D problems. The difference in the use of information between China and Sweden can be seen in the comparisons found in Table 1.

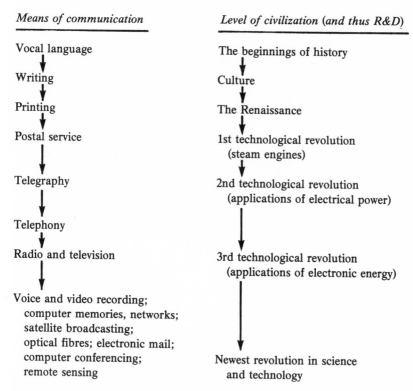

Means of communication	*Level of civilization (and thus R&D)*
Vocal language	The beginnings of history
Writing	Culture
Printing	The Renaissance
Postal service	1st technological revolution (steam engines)
Telegraphy	2nd technological revolution (applications of electrical power)
Telephony	
Radio and television	3rd technological revolution (applications of electronic energy)
Voice and video recording; computer memories, networks; satellite broadcasting; optical fibres; electronic mail; computer conferencing; remote sensing	Newest revolution in science and technology

Fig. 1. The development of communication, civilisation and the related R&D levels.

Table 1. Comparison of the functioning of two national information centres.

Information Service Tsinghua University Beijing (in fact, a technological institute)	Documentation Centre Royal Institute of Technology Stockholm (in fact, a centre for the exchange of data)
• 90 per cent service for (and all income from) universities	• 47.5 per cent service for (and income from) universities
• 9 per cent service for industry (the greater part for the large enterprises)	• 47.5 per cent service for industry (the greater part from small firms)
• 1 per cent for government	• 5 percent for government

The tradition in many developing countries (particularly China) is to ignore information, especially in the scientific and technological sphere, and this is an important reason for the underdevelopment of scientific technology in these countries. So it is time for a new concept, one that proposes that information—as fresh knowledge—is a kind of wealth, the same as money or materials.

The value of basic knowledge related to science, technology

A changing concept today concerns the value of basic knowledge. In Chinese history, fundamental knowledge related to natural science and engineering never merited high consideration. Now, along with the widespread diffusion of the products of scientific technology, high-technology products such as tape recorders, video cassette recorders and television receivers are gaining in popularity from the large cities of the Third World to its countryside. Not only do such goods change people's everyday life, they change people's value conceptions. People want to know how to choose, how to use, and how to maintain these modern devices. Although the existing ability to do so lies beyond the range of most people's current knowledge, the evolving situation incites people to learn more about the natural sciences and their technological applications.

New scientific-technological processes have even more of an impact on the knowledge of production. The profits earned by the application of new technics in agriculture and manufacture stimulate farmers and industrial workers to acquire new knowledge founded in science and technology.

A general knowledge base for R&D consists, as a matter of fact, not only in a basic knowledge of science and technology but also knowledge of one's national culture. The latter is extremely important for policy formulators, decision makers and managers because the adoption of new technologies should be based in part on such knowledge. 'Technology cannot be transferred, it can only be developed,' reminds Sweden's Sam Nilsson, director of the International Federation of Institutes for Advanced Study (IFIAS). The development (or adaptation) of new technology based on national culture will, at the same time, also effect a change on that national culture.

The individual general-knowledge base used by the researcher, decision maker or manager concerned with R&D can be summarized as follows.

- Mathematics (basic concepts, e.g. differential equations)

- Experience in one field of R&D (at least five years): necessary in order to understand the relationship between R&D and society

- Fundamental knowledge of the computer, e.g., basic concepts regarding algorithm languages

- Modern management methods, i.e., applied system analysis, matrix management methods

- Foreign-language knowledge (a reading knowledge of English is indispensable to anyone working in R&D)

- Basic familiarity with the social sciences (especially law, economics, one's national history, and the history of science)

The first and third elements above can be considered constituents of modern science and technology. The last is intimately related to culture, while the remaining three combine the sciences with the humanities. One's ability to observe, analyse, judge and conclude depends very much on such knowledge, and the popularization of such general information will enhance considerably the capacity for development of Third World nations.

Standards of living

In today's China even the peasants look upon the quality of life not only in the spiritual sense (loyalty, tradition, family harmony, and so forth) but also in a material way (food, clothing, shelter, television, tourism, even perhaps an automobile). The Chinese agricultural worker thus now compares his living conditions not only with those of the past and of his village neighbors, but with those of the rest of the world. This has never happened before in Chinese history.

Before 1980, the purchase of foreign merchandise was never really considered by ordinary Chinese people. Indeed, even if the peasant ever felt it necessary to acquire such, he had no possibility to buy. By 1984, a young peasant from a mountain village commented to me, 'Sony tape recorders have good quality. I want to buy a recorder only if it's a Sony.' This remark could be commented on in a political sense, that it conveys a significant historical change in the developing countries. But notions of efficiency, time, law and so on are also evolving via new products and processes emerging from scientific technology.

Development of institutions

The implications of new scientific and technological processes are not only relevant for the concepts and mechanics of the production system; they also affect the development of our institutions. These consequences can be seen from three main facets of the development of social institutions: the selection of experts, the decision-making process, and management.

Selection of experts

Experts play, and will continue to do so, an extremely important role in the development process. But what are the main criteria for their selection? Figure 2 summarizes these.

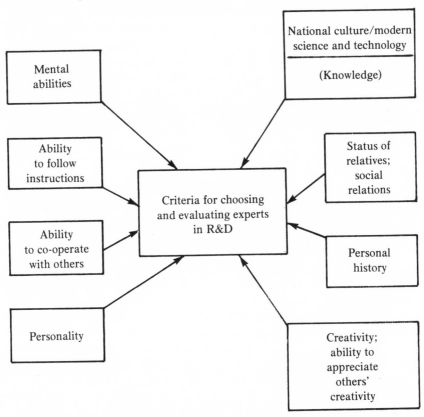

Fig. 2. Deriving criteria for the selection of R&D experts.

Which are the most important among these eight factors? Based on my own nineteen years in research and development, I would say that the most common answers in China before 1979 would have been 'status of relatives,' 'social relations,' as well as 'personal history.' Today, the most common replies would be 'knowledge' and 'creativity.' What brought about such a fundamental change? One of the main causes has been the advent of new science and technology: discovery and invention, designs, processes, products and services. During the course of the application of new processes or the use of new products, people come to realize that knowledge and creativity are the basic conditions for further research and technological applications.

Making decisions

In the late 1970s, decisions made in China concerning development were frequently the work of a few persons, operating on a limited basis and using random procedures. Now, a panoply of decision-making methods is used in China (e.g., applied system analysis). Figure 3 is my suggestion of the procedural scheme related to China's importation and adaptation of various modern technologies.

Numerous changes in China's decision-making processes are evident from my flow diagram. They can be summarized as follows.

- Decision makers are not a few individuals; they constitute groups including experts in any required field.

- The problem regarding which decision will be taken is considered, together with related factors, as a system.

- The decision-making process should follow certain procedures. Random decision-making is unsuited to modern science and technology.

- A variety of scientific methodologies (system analysis, feasibility studies, specialist forecasting) needs to be introduced.

- The feedback concept is extremely important as a corrective measure in the making of decisions.

The management function

The impact of new scientific technology is clearly felt in the management of the development process. Here, managers act not only as commanders

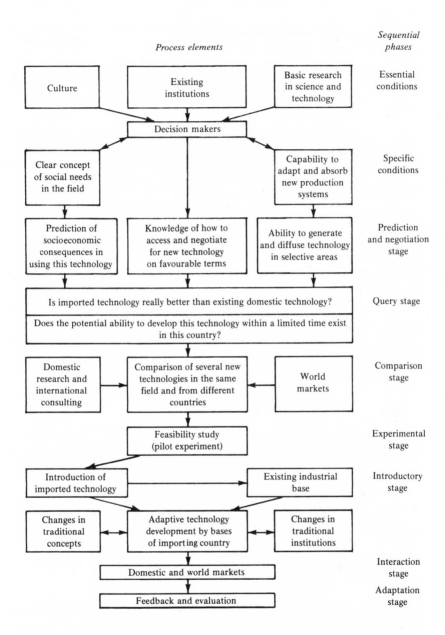

Fig. 3. Decision process concerning the import and adaptation of new technologies to the economy of China.

but as coordinators, staff consultants and service specialists. Their functional tasks may be summarized as in Figure 4.

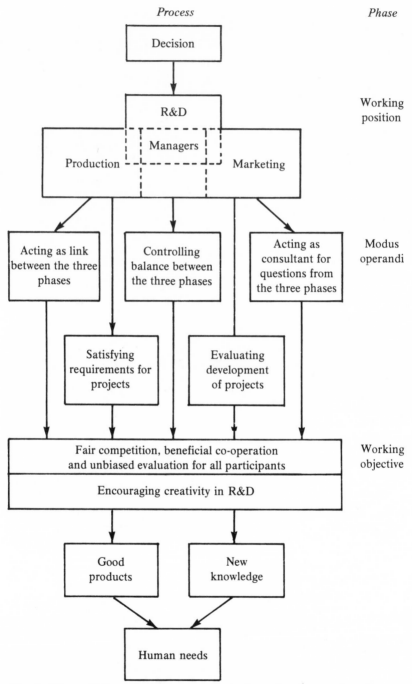

Fig. 4. What middle- and top managers do in the realm of research and development.

So the main role of the manager in the development of new scientific technology is to create and maintain conditions in which individuals, expert groups and integrated enterprises can compete fairly, work together beneficially, and evaluate objectively and justly.

The significance of individual creativity as the source of innovation in research and development cycles could be depicted as in Figure 5.

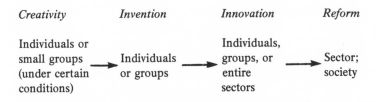

Fig. 5 The evolutionary pattern in the process of creativity.

It is clear from this simple schematic plotting of pattern that sustaining the creative role of individual experts is critical in the management of R&D, and that managers have a key role to play in this process. Managers have two imperative functions in this respect, as I have tried to depict in Figure 6.

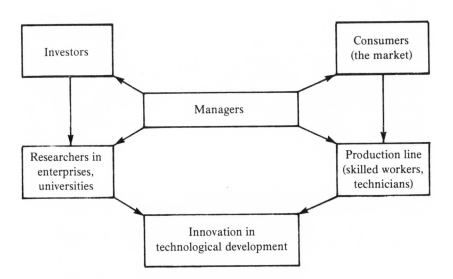

Fig. 6. Sources of motivation for innovation.

The first of these functions is the coordination of creativity originating in different sources, the four main sources being shown in the diagram to the right and left of 'Managers.'

The second responsibility is for managers to inhibit the creative output of one individual from limiting another's creative yield. Competition, on the other hand, can be introduced at any stage along the route of technological development. Large projects may be carried out, for example, by at least two groups having different ideas—even if one of the teams might be employing as little as a tenth of the total funds available.

The mobility of experts should be encouraged between diverse areas: from researcher to manager, between universities and industrial enterprises, between research institutes or 'think tanks,' or between disciplinary subjects (e.g., molecular biology and bio-engineering). Lifetime appointments generate enormous limitations on the creative spirit, especially in the domain of scientific and technological research.

By way of conclusion

Finally, one should ask: What is the most important factor as regards the advancement of scientific technology in contemporary China?

The most critical one, which is at the same time a condition and a consequence, is the continuation of the new 'open door' policy in China. China's development has been impeded for more than a thousand years by a mono-economy based on small-scale farming, and this development suffered in particular from the 'locked door' policy of the last two centuries. This situation was exacerbated during the past three decades by the elimination of competition.

The Chinese nation now feels keenly the need for a diversified economy, the continued implementation of its open-door policy, and the encouragement of competition through the influence of new scientific technology. Using this approach, the latent creative character of the Chinese people can play an ever increasing role in the contribution of innovation to the scientific and industrial world.   ■

A professor of chemical engineering examines the ways and means of leading students of science and technology through the maze of professional education to a creative, full and rewarding career. He questions seriously today's value systems in his sphere of interest, and weighs the relationship between social benefit and financial cost of our increasingly expensive teaching process. He calls on the industrialized nations to set a new example for the developing countries.

Chapter 22

The making of an applied researcher

Interview with Jean-Baptiste Donnet

A doctor in chemical engineering and former president of the University of Haute Alsace, J.-B. Donnet has spent a full career in training new students in his field, continuing his own research in applied chemistry, serving as a consultant to industry, and improving the management of education at the highest levels. A reserve colonel in the French Air Force, Professor Donnet has published (among numerous works) monographs and books dealing with practical research results in the field of solid-state chemistry. In 1987, the author was appointed to France's Conseil Supérieur de la Recherche et de la Technologie. Address: Directeur, Centre de Recherche sur la Physico-chimie des Surfaces Solides, 24 avenue du President Kennedy, 68200 Mulhouse (France).

1

What brought you into the world of investigative and inventive chemistry?

As a boy growing up in the mountains of central France, I was apprenticed to my father to learn how to make *sabots* (wooden shoes) and to become a hairdresser. I had a need to understand what I saw around me. This compulsion to learn, my mother later told me, first showed up when I learned to read by myself at the age of three. Then, when I was in a rural school run by a religious order, I pushed this curiosity to the point of mounting my own simple experiments to confirm what I was being taught in terms of elementary botany, physics and chemistry.

For example?

Well, we had been taught that water could be separated into its constituent elements by passing an electric current through it. But how would I do this in a mountain village? Because I knew that electricity feeds light bulbs, I broke open a bulb and used the electrodes of the filament in order to produce, very rudimentarily, the gases involved.

How old were you then?

I was fourteen. My findings fascinated me, but they could be dangerous: I succeeded in setting off a lovely explosion of hydrogen. The barber shop was next door, where my father was trimming a farmer's hair, and I had been following the description of an experiment in an elementary manual. The book said that hydrogen burns, so I struck a match—and Boom! Luckily, I wasn't hurt.

My second experiment required more time. This was to understand the mineralogical composition of my native Auvergne, rich in sulphur compounds (especially lead sulphide). It was fun, in that geological paradise, to see how I could decompose these compounds through chemical action.

Although I remained with my parents until the age of twenty, but without school certificates of any kind, I was able to accumulate quite a laboratory—developing my own X-ray source by way of a water siphon, first, and then a mercury siphon. A conclusion that I would like to draw here is that one tends to forget that the person who really wants to learn often finds the means immediately around him. I should add that I learned a good bit from the popular-science journals available at the time, especially *Science & Vie*. Their articles were well written and helped me go well beyond what I was learning in primary school, so that I had a real desire to take secondary-school courses by correspondence.

During the war, I was involved with the Resistance, and then I did my regular military service. By the age of twenty, I was able not only to pass the secondary-school leaving certificate (*baccalauréat*) but to be admitted, as well, to the chemistry institute in Strasbourg. I got a degree in chemical engineering, then had the opportunity to work up my doctoral thesis. Laboratory work was important for me, and it helped me to develop further a rigorous method of reasoning.

The sense of curiosity that you were aware of: did this come independently, or was there some sort of close influence, as in the case of Einstein's uncle?

This is the kind of question that is difficult to answer because the determining factors are hard to single out. Certainly my father urged me to learn gradually how to work with my hands as well as to seek to understand, but in a fairly basic way. My own perception of things was not to be satisfied with simple explanations.

2

This would seem to explain the reputation that you have now, one of an intellectual who does not mind getting his hands dirty.

Not only do I not mind, but I must remain close to natural phenomena, close to concrete reality. And my lack of gullibility extends even to the work done in my own laboratories. One of the things that never ceases to amaze me is how some intellectuals believe that all (or almost everything) has now been understood about the world we live in. A few years ago, while serving as president of the university here, I was asked the following question by a young man: 'Since you are a chemist, Sir, don't you think that we now know all about this discipline and that we should now direct our research efforts to the coming fields of biology and the computer?'

Here is how I respond to this kind of question. I believe very much that we stand at the dawn of knowledge. Our most evolved theories are still approximations—some of them very rough approximations, others somewhat less approximate than the ideas of Democritus or our own on the eve of the atomic age. We are still incapable of writing out, correctly and fully, the path of the simplest chemical reaction. For more complex reactions, we can approximate the process. When you think of the physical and chemical phenomena at work on surfaces (a field in which I have spent thirty years), I am thankful for the measuring instruments at our disposal; but we are still far from a satisfactory understanding of these phenomena.

In other words, the progression of your own knowledge has gone from the empirical and the world of handicrafts, to that of scientific experimentation and theory. Now, in your sixties, you are still unhappy about the progress you have made?

I think that the satisfied scientist is not a scientist at all, or else he is an astonishing creature. Let me take my own progression to explain what I mean. When I was doing my doctoral thesis, I had the luck of joining the large, multidisciplinary laboratory run by Professor Charles Sadron in Strasbourg. We were physicists and biologists as well as chemists, and the aim of my dissertation was to verify experimentally the validity of certain fall-out from the laws regulating brownian movement (established by Einstein in 1905).

These laws governing particle motion describe diffusion phenomena. The laws are simple, very easily verified by experiment, and they find vast application in technology as well as in research regarding transport phenomena: the transport of matter, of heat, of electricity. But what we tend to forget about these laws is that their approximate nature results from a hypothesis of the continuous medium in which they move—continuous by comparison with the size of the particles moving inside. This is what my director, Professor Sadron, wanted me to check by measuring (via electron microscopy) the dimensions involved. My job was then to establish experimentally the behavior of the particle models and to compare these results with the theoretical predictions. The dimensions of the particles had to be reduced constantly until I had arrived at the same order of magnitude as those of the molecules in the solvent of the dispersal agent.

Evidently, once the molecules of the dispersal medium and those being dispersed reach about the same dimensions, one cannot call the dispersal agent a continuous fluid in relation to the particles in motion. During the preparation of the thesis, my colleagues and I were very surprised by watching the behaviour of a continuous medium in a solvent. We did not follow through, but to this day the results continue to astonish.

Something like the relationship between quantum physics and quantum chemistry?

Stated otherwise, I believe that satisfaction in regard to science is not possible. Of course, one can be content knowing that one has done the utmost, but it is the residual dissatisfaction that prompts ceaseless new trials. This is a dialectic of scientific thought: on the one hand you are not satisfied, you want to go further; on the other hand and given a number of constraints, the results are completely satisfactory and you proceed to plan new trials in order to find the defects in your knowledge. I like this dialectic because it is a factor of progress, of acquiring new knowledge.

3

Have you known the truly happy researcher among your students and other disciples?

A good question, because the scientist who does not achieve some satisfaction would likely halt his work since this requires concerted effort. It is difficult to limit the intensity of endeavour to be always available, and to show limitless rigor. There are many satisfactions, of course, but I give priority to three. First (and I mention it yet again) is to understand better. Second is to be able to explain (understanding and explication are not exactly the same). Third is to realize that a prediction can come true—an immense satisfaction.

How about people who have achieved a certain level of understanding of nature and life, perhaps students who think they have learned enough to stop asking questions? Have you encountered any?

Absolutely, but this is not a question of age. The person, the student who thinks he need no longer ask himself questions attains a level of confidence that is tantamount to dogmatism. The course of human history is studded with people of this type: in philosophy, in politics and in science, although science is of course a form of philosophy. (I saw recently that my colleague, George Pimentel, addressing the American Academy of Arts and Sciences, said that in the future we should include chemistry, physical chemistry and the physical sciences in general within the category of philosophy.) To answer your question, I would say that the individual with a mental block or manifest self-satisfaction is neither comprehensible nor admissible as long as human knowledge remains so limited.

How would you explain, then, the attraction of the research world for a young man or young woman characterized by such self-satisfaction?

I am inclined to reply that this kind of phenomenon is both intriguing and infinitely dangerous. This is because of the changing proportion of the public interested in science and the realization, at last, that science, through technology, is an instrument of power. One must, naturally, distinguish technological applications from the scientific method itself. Yet it is incontestable that some of the big changes in our life stem from overtures in scientific knowledge. Indeed, it is for this reason that the western nations have invested so heavily in science education since the turn of the century; this is why so many youths have been drawn (if not naturally pulled) into scientific training.

I would say, furthermore, that a good many young people have no place in this adventure; and it is probably these who are the most susceptible to the self-satisfaction we discussed, susceptible even to disenchantment. I might add that it also follows that it is young people endowed with a good memory and with powers of analysis—rather than scientific curiosity—who have a good chance of succeeding in the school systems of the industrialized countries. These thus find themselves on the royal road to a career in one of the scientific disciplines, and this is a model being emulated in Third World countries. This helps explain why some young people, bright and with irrefutable cerebral power, still lack an inquisitive spirit and do not have that indispensable faculty for asking the right questions.

4

So, for you, the idea of progress subsumes a process of permanent re-assessment?

It seems to me that, if one has little desire to learn, the laboratory is not the appropriate workplace. Other than being a scrupulous worker attentive to detail relating to a particular field, one must ask all those questions that will open countless doors and windows on the mind. One will not always succeed, succeed perhaps rarely, and this is where modesty must reign.

In an increasingly interdisciplinary world, how do you and your colleagues envisage meeting the needs of cross-fertilization?

When I finished my thesis as a young man, I went to work for a year in a large laboratory in the Federal Republic of Germany where scientists from all the disciplines were thrown together. There, the fertilization of one's thoughts by the ideas of others became immediately evident. The simple exercises of description, comparison, query, and mutual satisfaction with my partners from other disciplines constantly bred new ideas. The situation is the same in my laboratory, here: we have working with us not only physicists and other chemists, but technologists and applied researchers as well. Thirty of our staff are working on goal-specific industrial contracts, and their interaction with the rest of us is continual.

There are also thirty-some foreigners coming from diverse countries, and this team renews itself constantly. Added to these are the young chemical engineers with bachelors' or masters' degrees, so that the totality of backgrounds is quite a mixture. One of my own satisfactions in this respect is that a complete diversity of backgrounds—other Europeans, Americans, North Africans, Lebanese, Syrians, Chinese and Japanese—are able to converge on common goals, discuss science and lifestyles (and even politics). And all this in two languages, because we speak mainly French and English.

So your research center is a cultural crossroads as well as a multi-disciplinary meeting-place?

Positively. We even play host to specialists from the Arab states and Israel who learn to live together, exchange ideas and respect each other, and take their meals together. Conviviality is essential. Laboratories are thus useful places to 'put together' tomorrow's world; there the scale of values is different, and there is no contest of force. The relationships have to do, instead, with competence, seriousness and honesty. In science, honesty is not negotiable: one is not a little honest, or very honest. There is either total honesty or none at all. Your research results have been determined with precision; you cannot bend them in order to fit the desired aims.

As a consequence, you begin your experiment by varying as much as possible the test conditions, resulting in the most convincing results that can be obtained. Experimentation is your master—your absolute master, and such discipline is primordial for young people coming from totally different cultural backgrounds. I believe that westerners are, in general, well adapted to such discipline.

5

Can we turn to some more abstract factors influencing progress in science? Individual freedom, for example . . .

The great jumps in our scientific knowledge, such as those made by Newton and Einstein, or the great leaps in technical advance made by inventors such as Edison, substantiate the thesis that innovation implies freedom: the freedom to be curious, and then to pursue the trail of discovery to its logical end. I do not visualize reaching the end of this trail without the greatest liberty possible. No one is ever fully 'free,' or without constraint; but restraining factors contribute in their own way, however, to mankind's progress. I am thinking of war, the horrors of war, that have wrought new applications of knowledge.

And yet one should make a distinction between levels of creativity, levels of ideas, levels of application. These are all different. In terms of application, constraints such as commercial restrictions could also be a factor of progress.

You're thinking of the demands of the marketplace?

Market demands, competition, cost/benefit ratios mean that a choice is to be made among the various options, that the various possibilities are tempting. But those susceptible to give in to the temptations are not really men and women of science. Our work cannot be founded only on the basis of compensation and reward.

As a chemical engineer, have you had the responsibility of creating analytical systems, from conception and implementation to evaluation? And in this regard, should economists show a more open mind towards the work of scientists?

To answer your first question, yes. The application of chemistry to the making of molecules depends on the use of the rigorous laws of thermodynamics. Despite the severe limitations imposed, thermodynamics has made it possible for the chemical industries to make fantastic progress, to the point that we now understand what we are doing in chemistry. We can forecast very satisfactorily how much such and such a product will cost, the best way of making the product, and what its properties will be. As to end-of-production analysis, we can purify to the extent required for medications calling for very few impurities in parts per million beyond which a product could have serious repercussions.

If the scientist makes it a point to work in such a manner, then it follows that the economist should be aware of the rigors of the approach and give the scientist full credit for meeting the criteria of an economist.

Given the nature of your reply, can the individual researcher be left a certain amount of time on his own to work uninhibitedly?

It is true that most research done today in industrialized countries is goal-oriented (and university researchers complain much about this). Truly pure research no longer finds the kind of support that it might deserve. Yet in the most competitive industries, in countries most open to competition (e.g., the United States), the really creative investigator will find the kind of support that will help him or her to develop professionally. I know that it is not easy to pursue this sort of individual research outside the organisation's budgetary framework—but it can be done, it is done, and I am glad of this. Keep in mind, however, that of the two million or so researchers active throughout the world today, a great proportion of these are followers rather than creators. They follow in the footsteps of their predecessors or of their companions.

Another thing: research today requires considerable technical means, all very costly. Research managers have to be careful to allot expensive equipment and instruments to those innovators capable of doing original work.

With the coming of big science almost two generations ago, laboratory-industry relations began to change. How has this been felt within the university research community?

The chemical industries being an extension of our knowledge of chemistry, they are clearly clients of what we know. The industry needs to learn increasingly about reaction pathways, processes and techniques for analysis and control, and product improvement. And today's chemical industry combines knowledge related to mechanical engineering and data processing. Basel, only thirty-some kilometres from Mulhouse, employs more people in its chemical industries than the entire French chemical sector. The Swiss have had a long tradition in this field of staying on top of the competitive possibilities in the pharmaceutical and agri-food businesses as well as in new chemical materials. The industry thus needs to maintain its own research momentum.

The industry calls, therefore, upon three kinds of research help. First, there are consultants, both individuals and groups: these are usually affiliated with university or governmental research. A second mode of assistance is to call on these same laboratories to contract for specific research at the expense of industry. (This has the added advantage of affording researchers the possibility of publishing their research results, when proprietary concerns are not overweening.) And a third means is permitting industry to send its own teams into universities or institutes where they will expand their knowledge and acquire new techniques. The Germans and the British developed these approaches first, and the United States followed immediately. The French and the Japanese are beginning to develop the habit. In the eastern European countries and in China, the academies of sciences and their research institutes collaborate with the industrial sectors as a matter of national industrial priority.

6

To what extent should the public expect scientists and engineers to take an active part in the formulation of policy and the making of decisions at the national level?

As far as national science policy (or research policy) is concerned, I must say that I know of no country where this is formulated without the participation of scientists. The whole process is a political one, however, and I don't see how it can be otherwise. It would be extremely interesting, nonetheless, to examine who the scientists are who take part in the process and just what they contribute. This would make a good theme for an international workshop.

I have taken part in the French process; I know that the scientists are consulted earnestly, and that their word carries weight. But money is the final determinant of first priorities, whereas national objectives (the country's ideology) are perhaps the second determinant. When President Kennedy decided to send men to the moon and when President Mitterrand launched the Eureka project, these were clearly the decisions of political leaders rather than of the scientific community.

If your question implies 'What role should the scientifically interested *public* play?' then I say that there must be such a role. For those of us raised in a democratic environment—perhaps not the best in the world, but I can imagine systems far worse—I do not see how I could say otherwise . . .

. . . Churchill said that until man devises the perfect system, democracy must remain the second best.

Indeed. The movement of 'public assessment of science' begun in the United States in the 1970s is intrinsically a good one, so long as it deters the making of decisions by small numbers of people (perhaps ill informed) in a large society. The process can deteriorate, however, if it is taken over by other ill informed people who convert it into a consumerist action or ecological militancy. I believe that most of us are seeking a good assessment process, an evaluation method that we might one day discover.

For the moment, I should add that in most countries (taking into account different cultures and political systems), two-thirds of the research activity results from decisions made by large enterprises, the transnational corporations and cartels, governmental mechanisms and the like. These decisions should dominate, for some time to come, research activity related to computers, aerospace, and the automotive and chemical industries. The decisions may not always be rational—but man's irrational behaviour is perhaps more important than we tend to believe.

What, in this respect, do you think should become the role of scientists and engineers concerning the further development of high-technology weapon systems?

To reply, let me backtrack a bit. I have said that one of my concerns is the role of the scientist in the process of public decision-making. I believe very much in this, despite the fact that scientists see their role somewhat fuzzily and often play this role poorly. This role is really a combination of information, training and personal courage. Because scientists exercise their prerogative somewhat inadequately in this respect, their posture in regard to armaments is even less well defined. Arms and man have been associated since the beginning of time, but this does not prevent me from saying that the diversion of scientific know-how to the production of new weapons is a travesty.

Having said this, I would then speculate on what might happen if engineers and scientists refused to participate in nuclear arms programs. My personal reaction would be total approval—but on condition that this apply everywhere. Otherwise those having the courage to disassociate themselves from armament activity would most likely find themselves soon overwhelmed by countries whose scientists did not exercise the same option.

What we need, therefore, is a continuing dialogue in order to find a common way, because it is an aberration that intelligent mankind contributes to mutual extermination or the imposition of destructive power. Man is by nature somewhat lazy, inclined to follow the lines of least resistance, so it behooves men of science to incite others to want to learn and to make it possible for political leaders to be as enlightened as possible.

What is the future of continuing education?

Continuing education is one of the great cultural innovations of our time. Is the process simple, however, and will it last?

Education has assumed almost disproportionate dimensions in modern society. In the industrialized nations, the total number of students attending classes—from the pre-school to the post-graduate levels—nearly equals that of the 'active' population—those who work regularly for a living. This is virtually a luxury for us. If we add to our educational bill all the costs of health, defense and leisure, I am convinced that we will soon find that we are paying for a system that we shall not be able to afford forever.

Such expenditures were possible, and justified, during the period of socioeconomic expansion that many countries experienced after the Second World War. But now that this rate has begun to slow, the burden of expenses is beginning to be felt everywhere. As we enter a prolonged period of zero- or near zero-growth of our economies (when some national performances will even reflect negative growth), the cost will no longer be tolerable.

It is time to question ourselves, therefore, on the validity of how much we are investing in education and training—on what the cost/benefit performance really is. I suspect that in due time we shall have to settle for a simplified approach to the challenge of educating a nation's youth.

7

Which returns us to the preparation of the next generation for the living of life. Is there a creative spirit inherent in the student?

This is a question asked, I suppose, by almost every teacher. I have no magic retort to the question, but I would have a few observations to make. The first is that when teaching is done well, it brings the student maximal information and a maximum of sources of data. Good teaching, that is, presents the essentials of learning, and then personal application and research on-the-spot will permit the student to learn more. This will depend very much on the teaching syllabus from the primary grades (now, even kindergarten) through university level and in the engineering schools. The curriculum should be attractive to the mind, varied, and designed to prevent boredom. Sitting on a hardwood bench is tedious, but most students have no alternative in order to get ahead in life.

I have the uneasy feeling, however, that sometimes the teaching system kills curiosity. Today's multimedia attractions outside the classroom, furthermore, add fuel to the fire. Besides, laboratory routines in university are often rather constricting, leading the student to believe that all has been accomplished already because everything has been tried, more or less. The net result of all this is that teachers very frequently no longer know how to discover and encourage the really creative student.

What can be done about this?

An idea that I have had for a long time (although I am not sure what it is really worth) is based on the good luck that I had—between the ages of fourteen and twenty—not to go near a university. I earned my living until the age of twenty and, by the time I was ready for professional training, I had an extraordinary and wild desire to learn. Once in engineering school, I worked day and night, and I learned a lot. Because I continued to believe that I knew only a small fraction of all that I wanted to know, I suppose that I have not really changed since.

But with youngsters who have been parked at their desks for years—I have seen this with my own children as well as with my students—there is a difficulty in driving oneself to learn. There are, of course, truly exceptional students who will not let themselves be 'deformed' by their teachers—some who will even tangle with their instructors. So we have here, as you can see, a basically difficult problem to resolve: School does not develop personality, it distorts it. And this is generally recognized. Some professors are even afraid of the student who asks the difficult questions.

Teaching the student how to use his free time is not the answer to the problem. I think that a much better idea is for the formal teaching process to help students find novel, useful activity. School hours (in France, at any rate) are too long. Teachers have to be, as is the case in the United States, for example, more diversified in ability, better prepared to help the student branch out, blossom, and help to develop himself into a well rounded human being. At the same time I admit that one cannot teach a trade to someone who hasn't the calling, nor can one educate someone who is not very interested. Perhaps the fundamental mistake that we teachers make is not to help young people understand that life expects from us as much as it gives. ■

Governments in the industrial economies have fashioned schemes to promote high-technology industries on the grounds that they drive and mould economic growth in a wide range of activities and that national competitiveness depends on strength in these industries. Japan does not give as much financial support for the development of advanced technologies, but it has gone further than western European and North American governments to assess the needs for such technologies in the 1990s and in cooperating with private firms on R&D programs designed to generate the necessary seeds.

Chapter 23

Seeds and needs—and other factors in innovation

Bruce R. Williams

Until 1986 Professor Sir Bruce Williams, KBE, directed The Technical Change Centre, 114 Cromwell Road, London SW7 4ES (Great Britain). He is an economist trained at the Universities of Melbourne, Adelaide and Manchester who served as vice-chancellor and principal of the University of Sydney from 1967 until 1981. He edited The Sociological Review *(1953-1959) and is the author of* Investment in Innovation *(1958).* Investment, Technology and Growth *(1967),* Science and Technology in Economic Growth *(1973),* Living with Technology *(1982) and* Knowns and Unknowns in Technical Change *(1984).*

Early philosophical development

In the first significant treatment of economic growth through technical change, *An Inquiry into the Nature of Causes of the Wealth of Nations* (1776), Adam Smith took the definite view that in the interests of technical change, government activity should be severely restricted. Seventy years later, in *The Communist Manifesto* (1848), Marx and Engels wrote in the most glowing terms of the power of competitive market systems to generate technical change and growth. But unlike Smith, they thought that technical change would produce such economies of scale that competitive pressures to innovate would be progressively diminished and the growth of a world of monopolies would make the market system too unstable to survive. Almost 100 years later, in *Capitalism, Socialism and Democracy* (1943), Joseph Schumpeter argued that the growth in the scale and power of organised R&D was turning innovation into a routine function in large corporations (i.e. Marx's monopolies).

This growth in the scale and power of R&D was made possible by a substantial non-market provision for education and research. Even before the First World War, it was clear that Germany and the USA were exploiting the opportunities for innovation in the new science-based electrical and organic chemical industries more effectively than Great Britain. This was because of their greater provision for the education of scientists, engineers and technicians, and scientific research in universities. The growth in the importance of industries which depended on research and on the recruitment of graduates for both military production and competitiveness in international trade, the high proportion of applied R&D conducted by large firms, and the increase in the proportion of work directly organised by managers in medium-sized and large firms, carried with them a greater role for non-market forces in economic growth than Adam Smith and his followers envisaged.

After the Second World War, governments in the industrial countries adopted more positive policies towards the expenditure of public funds on education and on R&D to promote technical change and growth. There was a marked growth of research into the ways in which investment in education and research promote growth, and into the nature and extent of 'externalities' which cause reliance on market mechanisms to produce sub-optimal levels of investment in research. But the literature in this field was more influential in creating a climate of opinion in favour of an increased role of government than helpful in establishing how much governments should spend, in what areas, and in what manner. The sharp reduction in growth rates from the early 1970s and the consequential increase in the budgetary problems of governments encouraged re-appraisals of their role. One significant sign of this re-appraisal was contained in an OECD report, *Technical Change and Economic Policy* (1980), that advocated less financial involvement of governments, or government agencies not involved in production, in development and design work near the applications stage, and more in research and exploratory development in fields likely to have potentialities for innovation across a range of activities.

The researches of Crick and Watson which revealed the chemical structure of DNA in 1953, transformed the productivity of research in biochemistry and related fields, and led 20 years later to developments in genetic engineering with potentially great applications to agriculture and industry. A considerable number of scientists with the relevant skills were persuaded to become directors or consultants of companies wishing to use the new knowledge and techniques, and some scientists formed their own companies. The close link between the dates of citing bioscience papers and biotechnology patents led some observers to argue that science and technology are merging.

This rapid take-up of the new biotechnologies, and a somewhat similar experience in some fields of electronics, has persuaded some governments that universities should get more directly involved in innovation, and the expectation of vast profits—successful patents or production ventures—has persuaded some universities to reach the same conclusion. There is, however, good reason to be cautious. The evidence does not as yet justify a generalization that science and technology are merging, patents as such do not constitute technology which requires production, the risks of innovation are still substantial, and the directness of the links between university bio-science laboratories and innovations in processes and products (of which there have been very few as yet) may soon decline, as it did in the electrical, organic chemical, and computer industries as success in innovation made possible great increases in production and the development of in-house facilities for R&D.

Smith and Marx

Adam Smith put great emphasis on the role of division of labor in innovation and the growth of wealth: as workers specialize in particular operations, they invent easier and readier methods; as making machines becomes a specialized occupation, the makers improve them; as science becomes the principal or sole occupation of a particular class of citizens, scientists improve and invent machines. Regulations restricting trade and the nature of production activities therefore impeded innovation and growth. In a 'system of natural liberty,' individuals would be keen to better their own condition, competition between producers would provide pressures to innovate, and price movements would coordinate the activities of individual producers and consumers.

Smith's theory, which still influences American views on the government's role in technological development, gave non-market factors a rather minor role. The government would need to provide for defense, the administration of justice, and those public works or public institutions which though socially advantageous would not be financed by private enterprises. What he defined as desirable public works and institutions were those roads and bridges that could not be financed from tolls, facilities abroad to facilitate commerce, and institutions for the instruction of the people, to provide basic education at less than cost for children of the poor, and paradoxically, for adults to offset the tendency of a fine division of labour to induce stupidity, ignorance and lack of energy in private affairs, and rash and capricious judgments in public affairs.[1]

By the time Marx (with Engels) wrote *The Communist Manifesto*, the collection of groups of workers in factories was well established in Britain, strikes by discontented factory workers provided a new incentive to invent labour-saving machinery,[2] and trade cycles and periodic financial crises had become normal features of the market system. In *Grundrisse* (1858) Marx wrote that in advanced heavy industry all the sciences had been pressed into the service of capital and invention turned into a business.[3] The capacity to harness science to the service of capital and to accumulate capital and invest it in innovations, explained why capitalism had been able to create during its rule of science for one hundred years more massive and more colossal productive forces than all preceding generations together.[4] 'Social labour' in factories was more productive than individual labour, but the capitalist form of social labour was distorted by the power of capitalists to force workers with no means of production of their own to become mere appendages to machines. By contributing to the discontent of workers and so stimulating investment in labour-saving inventions, this distortion would contribute to the downfall of capitalism. For it would increase technological unemployment, and via a rise in fixed capital relative to circulating capital add to the tendency for the rate of profit to fall. Marx also predicted that technical change would add to the economies of scale, and lead through increasing monopolization to greater instability in the market system. Because he regarded the capitalist system as both transitory and unsatisfactory, Marx did not consider the steps that might be taken, for example by anti-monopoly legislation, to prevent the progressive increase in the direct organization of productive activities and the diminution of the role of coordination by the price system. But his analysis implied that the growth of the former at the expense of the latter would destroy the capacity of the capitalist system to go on 'constantly revolutionizing the instruments of production.'

Marx had little to say about differences between inventions and innovations, and had little to say about the effects of the creation of new industries on the degree of monopolization and labour skills. In his *Theory of Economic Development* (1912), Schumpeter emphasized the financial risks involved in using inventions to make process or production innovations, the role of entrepreneurs in taking these risks and in making the managerial and marketing innovations required to complement the technical innovations, and the reasons why innovations and their diffusion created trade cycles.[5] At first Schumpeter treated invention as exogenous, but in 1928 he wrote that planned inventions were making large corporations the main vehicles for technological innovation,[6] and then in 1943 he predicted that the entrepreneurial function would decay as organized R&D made it possible for large corporations to produce innovations to order.[7]

Although there are as yet few signs that innovation has become a routine function in industry, the growth in the role of industrial R&D has brought considerable change in the balance between market and non-market factors. Both Germany and the United States exploited the opportunities for innovation in the new electrical and organic-chemical industries (which were much more 'science-based' than older established industries) better than Britain because they had made much more (non-market) provision for scientific, technical, and technological education.[8] Japan's high level of education has also been a major factor in its very rapid growth rate since the Second World War, and in its more recent success in high technology industries.

Enquiries into the role of R&D in economic growth grew rapidly after the Second World War, and there is now a large literature on 'the residual factor' in economic growth—namely, the growth that cannot be explained by purely quantitative increases in capital and labour inputs. These enquiries led to a greater emphasis on the role of public expenditure on education, on basic research that it would not pay even very large corporations to finance but which would sustain the profitability of applied R&D in industry, and on applied R&D in small-firm industries, including agriculture and medicine.

It is now part of the conventional wisdom that non-market factors are of considerable and probably growing importance in innovation, and the main debate is about the extent of public expenditure, and about where in the range of activities from basic research, through applied research, development and design work, to capital expenditure and marketing, government actions are most likely to be effective in promoting innovations that will contribute to growth and to strength in international trade; also—even in some cases in fields of military R&D—about types of R&D which, though requiring public finance, would be best not carried out in government institutions. This debate explains the keen interest in the well documented Japanese debate, and the decisions on the programs and procedures for base technologies expected to be needed by the 1990s.

Japan's 'seeds and needs' approach

In 1981 the Japanese government launched a ten-year program for the development of the 'next generation base technologies'—defined as the technologies likely to have widespread industrial applications in the 1990s. As part of the process for establishing this program, a very large number of 'industrial problems demanding solution through research' were classified by field and mapped into a seeds (*seedzu*) and needs (*needzu*) matrix. There were three seeds columns—miniaturization, informatics and composite agglomeration—and three needs columns—individual human needs, overcoming the constraints of energy and resources, and overcoming the constraints of physical space.[9] Twelve broad projects were chosen for cooperative research in government laboratories and the research departments of private firms. These projects (six in the field of new materials, three in biotechnology and three in electronics) were judged to justify government initiatives and financial assistance because of the long 'lead times' involved, the large financial risks, and the potentialities for applications in several different industries.

It is unlikely that this matrix played a decisive role in the final choice of projects for the 'next generation base technologies' program. MITI's Agency for Industrial Science and Technology (AIST) has other larger programs such as the Research and Development Program started in 1966 (which currently includes projects on high-speed computer systems, interoperable data-base systems and advanced robot technology), the Sunshine Project (new energy technology) started in 1974, and the Moonlight Project (energy conservation technology) started in 1978, as well as smaller projects on medical and welfare equipment technology and regional technology.[10] However, the matrix approach helps to ensure attention to the dynamics of markets as well as of technologies in the decision processes of MITI.

Until recently there was little temptation for MITI or for private industry to finance R&D without a careful consideration of short-term market prospects. Japanese firms were able to achieve rapid growth by devoting more effort to world technology reconnaissance and the purchase of licenses to use western technology than to their own original research, and by developing high standards of production engineering and company organisation. But during the 1970s there was a major shift in Japan's attitude to indigenous research. This was a consequence of growing environmental problems, the fears created by the sharp rise in oil prices that Japan would face growing difficulties in sustaining exports from energy-intensive industries, the continued growth of electronics despite the world recession, and concern that success in absorbing and improving western technology would soon reduce Japan's capacity for rapid growth. Overcoming the constraints of energy and resources—Column 2 in the needs/seeds matrix—became an important component of 'needs,' and this, together with the prospects of diminishing returns to reliance on adopting and adapting western technology, showed the need to 'seed' the creative and autonomous development of new technology.

There is an old saw that necessity is the mother of invention. In creating its 'next generation base technologies program,' the Japanese government's MITI in effect extended the concept of necessity to include prospective necessity.

Except from time to time in defense where the military planners constitute the market, governments and their agencies do not have a good record in forecasting needs or in generating the appropriate technological seeds, and it will be some time before the success or otherwise of Japan's next-generation base-technologies program will be known.

Dore gives four reasons for expecting that the exercise will prove successful:

- Japan's world technology renaissance capacity is very considerable,

- as a function of nationalism and a sense of the need to catch up, information is more readily shared in Japan than in most other countries, primarily at meetings and conferences organized by think-tanks and government and in industry associations where the guild tradition of cooperation between competitors is still strong,

- the consensus nature of Japanese society and the high level of education make it relatively easy for staff in MITI to prepare *Visions* for the future from many sectoral analyses and judgments of trends and prospects in markets and technologies, and

- the matrix organisation of MITI—divided partly into industrial sector units, partly into functional units such as economic policy, overseas trade and technology promotion.[11]

However, even if their forecasts of scientific and technological potentialities and market prospects prove to have been correct, and their research programs prosper, it does not follow that Japan will be successful in meeting needs. Before even the best seeds can provide for needs, they must be sown and well cultivated. During the last 40 years governments in the mature industrial economies have spent substantial sums to generate new technologies, but there have been many examples of failure due to poor planting out and ineffective cultivation. Governments in Great Britain have a particularly unenviable record of failure. Japan, which has already shown a good capacity to adopt and improve established innovations, has yet to demonstrate that it can be a successful primary innovator. Although Japan's growth rate is likely to taper off with the closure of the gap between Japanese and U.S. technology and incomes,[12] few of her competitors would rule out the possibility that Japan will extend its capacity for primary innovation in, e.g., information technology and biotechnology.

If Japan does produce good seeds, three factors in particular will help in their successful cultivation. The first is restriction of cooperative research between government laboratories and private firms to the generic stage of research, leaving the private firms to finance and conduct the detailed development and design work involved in creating new or improved products and processes. In Europe many failures have been due to the involvement of governments, or of government agencies not responsible for commercial exploitation, in detailed development and design work. The second factor is that Japan has a very good supply of competent engineers and skilled process workers who are able to cope with many of the unforeseen problems that come with most innovations.[13] The third factor, which follows from the nature and high level of participation in education in Japan, is that the widespread involvement in the formulation and discussion of the needs and seeds program has created a climate of opinion and expectation likely to favour the adoption and improvement of the latest technologies in a wide range of firms and industries.

Until recently, few commentators thought that the Japanese government's approach to innovation had any relevance to problems in the United States and western Europe. But the continuing growth of the Japanese economy despite the world depression, the significance of Japanese innovations in established industries such as vehicles and machine tools, and the growing importance of Japanese firms in high-technology industries such as electronics and in certain areas of biotechnology, persuaded many commentators that at least some of Japan's policies and procedures might be exportable. The study of these policies and procedures has contributed to a reappraisal of the respective contributions to innovation of market and non-market factors, and of competitive and non-competitive forces. A good example is Nelson's comparison of high-technology policies in the USA, France, Germany, Japan and the United Kingdom.[14]

Non-market factors in the USA

Despite the free-enterprise rhetoric in the USA, the American government has contributed significantly to technical change and growth. From the 1850s, it assumed a major responsibility for training in the agricultural and mechanical arts, and later in the century the federal Department of Agriculture started R&D programs for agriculture. Public institutions have provided virtually all the secondary education and the bulk of higher education, and after the Second World War the federal government assumed responsibility for the support of scientific and technical education and university research. Since the Second World War, at least the percentage of GNP devoted to R&D has been higher in the USA than elsewhere, and between 1960 and 1980 a little over 40 percent was financed by government.

The experience of the First World War convinced the USA and other governments that certain advanced technology industries were important for national security. Between the two World Wars, the U.S. government ordered new military aircraft and helped to strengthen the technological capacity of firms making airframes and aero engines, and encouraged the formation of the Pratt and Whitney aircraft engine company. It also encouraged the growth of the radio industry, and the Radio Corporation of America was formed, under governmental prodding, to increase American strength in radio technology and to cut through certain tangles about patents; the express purpose was to push [the] industry to the forefront of radio technology.'[15]

After the Second World War, U.S. governmental aid for high-technology industries became more general and systematic. The National

Science Foundation was established to encourage and support the training of scientists and engineers and to finance basic research in universities, the Department of Defense and later NASA financed R&D in aircraft, engines and electronics and contributed substantially to the U.S. dominance in electronic computers, semiconductors and commercial jet aircraft, and the Atomic Energy Commission sponsored the development of nuclear power stations. Since the Second World War the diffusion of technical sophistication outwards from military science and technology into the civil economy has grown in importance, even for new technologies less obviously related to military applications. Medical developments such as antibiotics, techniques of blood preservation, and the use of chemical pesticides to control disease vectors were initially introduced in connection with the military.[16]

Between 1960 and 1980, just over one-third of the R&D in the USA was classified as military. Only a small part of military support for advanced technology was designed to affect the civilian economy. Neither the Department of Defense which financed most of the exploratory R&D that led to the early electronic computers nor the companies involved expected a large civilian market to develop, though in their current 'very high speed integrated circuit' (VHSIC) program and in other programs aimed at enhancing computer design capability, the Department of Defense is concerned with strengthening the capacity of American companies to get or stay ahead of foreign firms in electronic technologies.[17]

Non-market factors have been of considerable importance in the USA, though the extent to which government R&D programs have been contracted out to the market sector has reduced the extent of detachment from market influences. In 1983, 16 U.S. companies which make semiconductors and computers established the Microelectronics and Computer Corporation, and another group of firms the Semiconductor Research Corporation, for cooperative research. To ensure that anti-trust legislation did not impede these cooperative arrangements, in 1984 the Congress passed the Joint Research Management Act to validate cooperation in generic R&D by otherwise competitor companies.

Such cooperative arrangements for research ensure attention to market factors rather more than research contracts with universities or government laboratories might do, but they extend further the extent of the administered coordination of production activities.

In Europe, there has also been an extension of such cooperative arrangements. In 1982 the British Department of Trade and Industry introduced a national program for advanced information technology on the lines of Japan's fifth-generation computer project. This involves collaborative research between electronics firms and university and government laboratories before the commercial design and development state, and is financed jointly by government and industry. In 1983 the European Commission launched ESPRIT, a similar project involving 12 European companies and universities. The reasons for this growth in cooperative arrangements are the high cost of the range of research thought to be necessary to ensure success, the importance of the leading-edge technologies to national strength in international trade, and the suspicion that technological leads will prove to be cumulative. But, whatever the reasons, government-organised or validated arrangements indicate a judgment that it is not in the national interest simply to leave invention and innovation to the market.

Are science and technology merging?

A needs-and-seeds approach does not necessarily imply that all invention and innovation can be planned. Research programs derived from an assessment of needs may not generate highly productive seeds, while research based in the purely scientific interest in the chosen problems may create the seeds of a solution to needs by a different route, or create seeds which generate new needs. Although Sir Frederick Soddy, who worked with Lord Rutherford on radioactive decay, envisaged energy released in radioactive processes making 'the whole world a smiling Garden of Eden,' to the time of his death in 1937 Rutherford dismissed talk of nuclear power as moonshine. Watson's and Crick's discovery of the structure of DNA did not derive from a seeds-and-needs matrix, and their research plan was treated with considerable skepticism by other scientists.[18] The inventors of the first electronic computers did not envisage their further development for commercial data-processing, nor for the development of consumer 'needs' for personal and home computers.

The great increases in the amount of R&D and in the power of scientific instruments have made it more possible than in the past to invent to order, but there are still major uncertainties in the results from both research and investment in new products and new processes of production. These uncertainties are relevant to the pressure on universities in several countries (though not, it seems, in Japan) to pay much closer attention to national needs (other than excellence in research regardless of the field of science), and to include the application of the results of research to real-life problems in their objectives.

Much university research in, e.g., faculties of medicine, veterinary science, agriculture and engineering, is concerned with solving real-life problems, and members of the faculties of medicine, veterinary science and agriculture are able to develop and try out the most promising results of their laboratory research in their teaching hospitals, clinics and experimental farms. These opportunities do not exist in the other faculties—or have not in the past—but the number of patents held by academic scientists in the fields of biotechnology and electronics, and the industrial demand for the services of academic scientists in these fields, have persuaded some observers that the opportunities for application will become more general. That is, the gaps between science and technology, and between invention and innovation, have closed to such a degree that universities can be expected to play a very much more dominant and direct role in innovation than hitherto. Britain's government, for example, has made it clear that it expects university research to become more industrially relevant; it is encouraging universities, and their staffs, to patent their discoveries and play an active part in industrial applications.

Some universities in the USA and Europe have participated in the formation of companies to exploit their inventions, and a small number have set up their own venture-capital firms. The quantitative significance of such business activities, however, is very small and likely to remain so. The risks involved are considerable, as can be seen from the fate of many new electronics companies and from the (as yet) minuscule financial returns on the large investments made in genetic engineering. It takes much more than skills in research to produce commercial winners, and few patents repay the costs of patenting.

There is a further limitation, namely, that the dependence of industry on the knowledge and skills of the academic scientists and engineers who developed a new field which had important industrial applications does not last. The successful firms develop their own research, and the role of university science departments reverts to their normal role of making indirect contributions to innovation through the provision of graduates, scientific publications and consultancies. Thus university scientists and engineers played an important part in the design of the first electronic computers, but the industry now has its own very impressive research capacity, and it is often necessary for universities to appoint visiting professors from computer firms to keep pace with development in design. If genetic engineering has even a little of the industrial significance currently expected, it will not be long before the successful firms (some of which will doubtless be chemical and pharmaceutical companies that already have a large research capacity in complementary fields) develop a substantial in-house capacity.

Innovation seen and unseen

The planning of R&D to achieve specific objectives has grown in importance. Where R&D is meant to contribute to innovation and economic growth, it is important to adopt a seeds-and-needs approach: on the one hand, to increase the likelihood that the outcomes of the chosen R&D programs do have market potential and, on the other hand, to maintain interest in the potential of R&D to create new or improved processes and products in order to improve incomes and competitive strength.

But there will continue to be the possibility of important discoveries, and quite unforeseen discoveries, in science and engineering that a tight needs/seeds matrix approach might prevent. Non-market factors will be at least as important in the future as in the past. Research in universities has made significant contributions to those non-market factors which have added to the innovation process. Measures to push universities into more than a marginally direct role in industrial innovations will have a retarding—not an accelerating—effect on economic growth.

Although it is not possible to specify the optimal distribution of university research between projects that derive from 'real life needs' and projects that derive from the internal criteria of purely scientific interest, it is important to maintain excellence in basic science as well as skills in engineering in the definition of a country's needs.   ■

Notes

1. *The Wealth of Nations,* Book V, Chapter 1, Part III.
2. In response to an appeal from cotton manufacturers when their spinners went on strike, Richard Roberts invented the self-acting mule. Andrew Ure described this invention in his *Philosophy of Manufacturers* (1835) and commented that 'when capital enlists science in her services the refractory hand of labour will always be taught docility.' Marx wrote that after each new strike of any importance there appeared a new machine. In The Directions of Technological Change: Inducement Mechanisms and Focussing Devices, *Economic Development and Cultural Change,* October 1969, Rosenberg quotes several examples of inventions induced by strikes between 1820 and 1860.
3. *Grundrisse,* London, Allen Lane, 1973, p. 572.
4. *The Communist Manifesto,* Penguin Books, 1967, p. 85.
5. In *Business Cycles* (McGraw-Hill, 1939), Schumpeter extended his theory of the cyclical nature of the processes of innovation and diffusion to cover, in addition to trade cycles of 9-10 years, short cycles of about 40 months and long cycles of 40-50 years, although the evidence for the existence of long cycles is far from conclusive.

6. The Instability of Capitalism, *The Economic Journal*, 1928, pp. 361-86.
7. *Capitalism, Socialism and Democracy*, New York, Harper and Row, 1943.
8. See R.R. Locke, *The End of Practical Man*, Greenwich, Conn., JAI Press, 1984.
9. See Ronald Dore, *A Case Study of Technology Forecasting in Japan, The Next Generation Base Technologies Development Programme*, The Technical Change Centre, London, 1983.
10. See *AIST*, 1985, (Ministry of International Trade and Industry, Tokyo) for budget allocations in 1985.
11. Dore, *op. cit.*, pp. 23-25.
12. In 1980 production per worker-hour in Japan was only about 55 percent of that in the USA. However, because hours of work per year in Japan were 20 percent higher than in the USA, Japanese production per worker-year was 70 percent of the American.
13. For the importance of the capacity of firms to concentrate the attention of competent staff on problem innovations, see Bruce R. Williams and W.P. Scott, *Investment Proposals and Decisions*, London, Allen and Unwin, 1965.
14. R.R. Nelson, *High Technology Policies, A Five-Nation Comparison*, Washington and London, American Enterprise Institute for Public Policy Research, 1984.
15. Nelson, *op. cit.*, pp. 31-2.
16. Harvey Brooks, Science Policy and Commercial Innovation, *The Bridge*, Summer 1985, pp. 7-8.
17. Nelson, *op. cit.*, p. 45.
18. See *The Double Helix* of J.D. Watson, London, Weidenfeld and Nicholson, 1968.

PART V

Holism and Change in Creativity and Invention

Some historical streams of progress in philosophy, science and technology are described. Imagination, intuition, ability, coincidence, opportunity and sometimes luck have roles to play in the drama of innovation. Reflection and the process of 'thinking through' have been vital to the development of philosophy, the natural sciences, and technology based on scientific knowledge. Experience combined with opportunity has had an unusually important part to play in the emergence of some kinds of social innovation. Here are treated examples not cited elsewhere in this book, and the main psychosociological pressures lying at the roots of creativity and invention are identified.

Chapter 24

The historic boundlessness of the creative spirit

Jacques G. Richardson

The present consists 90 per cent of the past.

Fernand Braudel

More than five centuries of evolution in philosophic thought

The era of Copernicus and company (see p. 3 of Chapter 1) saw the emergence of novel philosophies geared to new trends in social institutions, such as the banks grown from Florentine, Genoese and Venetian trading habits, the centers of navigation and cosmography (world maps) in Portugal, Spain and some culturally related cities of the Arab world, as well as from the phenomenon of the nation-state itself.

John Wyclif of Oxford, repulsed by the authority and corruption rampant in one institution, the Roman church, sought the church's secularization (in 1379). His efforts failed, but they set the tone and led the way a century and a half later to the Reformation, the general renovation of a prime western theological system that had been solidified a thousand years earlier by an Algerian convert, St. Augustine, 'a major intellectual, spiritual and cultural force.'[1]

Sir Francis Bacon, a true son of the aftermath of the Middle Ages, made two cogent observations in his *Essays* about the innovative processes. 'As the births of all living creatures at first are ill-shapen,' he noted, 'so are all innovations, which are the births of time.' Bacon also cautioned that 'he that will not apply new remedies must expect new ills,' whereupon he repeated that 'time is the greatest innovator.' Time and creativity can thus be considered handmaidens.

Progress is a cumulative procedure, as we realize now, that absolves the individual as well as the collective memory from returning each time to some indeterminable stage of departure. We build, rather (and luckily, too), on those who went before and on their total endeavour. After the contributions of Wyclif and Bacon, for example, we can trace selectively the main evolution of much modern western philosophic thought—and here we are required to adopt a clearly Eurocentric outlook. Table 1 traces the main lines of western philosophic evolution after the Middle Ages were left behind and makes the point of the accrual of knowledge.

It is reasonably clear from this summary of inventive philosophic and social thought, stretching over more than a half-millennium, that 'scientific analysis' gradually gave way to new understanding of the human condition. Theory, that is to say, was summoned to help understand practice; and, ultimately, practice was made the servant of hypothetical analysis—both evidently being needed to help advance the cause of a humane understanding of humans and their culture and civilization.

Is the understanding of innovation a discovery of reality? Is it a true interpretation of the ways of progress? Yes . . . and no. Bacon and Descartes tried to sort out from the curious tangle of 'social advance'

Table 1. Evolutionary depiction of western reflective philosophy subsequent to the Middle Ages and the Crusades. Note the 600-year cycle implicit in the individual emphasis made by Wyclif and Sartre.

Innovator	Time	Contribution
Wyclif	1320-1384	'Each . . . shall be saved by his own merit'
Bacon	1561-1626	Experiment, inductive method
Hobbes	1588-1679	Mechanistic materialism
Descartes	1596-1650	Linking of arithmetic-geometry; refined reductionism
Pascal	1623-1662	Computation of probability*
Huygens	1629-1695	
Spinoza	1632-1677	(May be too marginal for listing)
Locke	1632-1704	'New way of ideas': a theory of knowledge
Newton	1642-1727	Enormous influence: universal laws of motion, inverse-square forces**
Leibniz	1646-1716	As Newton did, devised calculus
Hume	1711-1776	Critique of metaphysical dogma
Rousseau	1712-1778	Lyrical return to nature
Kant	1724-1804	Attempt to transcend critique of metaphysical dogma
Jefferson	1745-1826	Declaration of Independence
Bentham	1748-1832	Utilitarianism
Saint-Simon	1760-1825	Positivism; precursor of social science
Fichte	1762-1814	Dialectical 'absolute idealism'
Hegel	1770-1831	Reconciliation of history with reason; idealism
Schelling	1775-1854	Return to 'absolutes'
Biedermann, Fischer	19th century	'Orthodox Hegelianism'
Stirner, Marx (who defected)	19th century	'Left Hegelianism'
Strauss, Bauer	19th century	Historical criticism, 'right Hegelianism'
Kierkegaard	1813-1855	Genesis of existentialism
Engels	1820-1895	Critique of 'positive materialism'; dialectics
Peirce	1839-1914	Pragmatism
Croce, et al.	Turn of 20th century	Neo-Hegelianism

[Here enters the 'non-bourgeois socialism' of Liebeknecht, Bebel, Kautsky]

Wittgenstein	1889-1951	Positivistic mysticism
Sartre	1905-1980	Contemporary existentialism

*Pascal's strongest influence concerned religion; Huygens' impact on philosophers was little.
**Newton is particular, and some specialists may prefer to omit his name.

what might be coherently constructible by the logical mind. Their successors brought the element of quantification into the picture slowly (the Pascal-to-Leibniz accretion), and *their* followers gradually gave way to qualitative interpretation of the real value of progress in its societal manifestations.

Marx, who in the last century contributed to theoretical innovation, refuted or at least failed to grasp the work of a great contemporaneous innovator, Darwin, whose ideas on evolution (which Darwin himself chose to call 'descent with modification') Marx branded the *bellum omnium contra omnes* or the war involving all against all. The seeming randomness inherent in fitness, survival and evolution would have to wait its holistic rationalization, well into the 20th century, by the likes of Ludwig von Bertalanffy (general theory of systems), Richard Goldschmidt, E.H. Carr, the Eldredge-Gould-Vrba trio (and a good many more): the combination of general system theory and the modified evolutionary concept of 'punctuated equilibrium.'[2] Punctuated equilibrium views mutation in genetic spurts, instead of a smooth process of step-by-step transformation of morphology, function and behaviour.

Historical shifts in the centres of innovation

We have just traced, admittedly in cursory form, how reflection affected 'natural philosophy,' or how the human sciences were translated into the 'exact sciences' from the era of the Italian Renaissance until now, when the industrial revolution is well into its third century. This has coincided with a period of extensive exploration and discovery (in their geographical notions), when our world was sensed as a sphere by most thinking people, when the solar system became acceptably heliocentric, when astrology became astronomy, alchemy became chemistry and Locke's 'new ways of ideas' gained growing respectability in many cultures.

In other domains, the city-state and then the nation-state became fundamental jurisdictions that prevail to our time in matters of how *res publica* is constituted, ordered, administered, taught, financed and defended. The old regions of Araby, Africa and the Americas waned before new empires waxed in the same areas. Literacy spread (although not fast enought), communication and transport facilitated both commercial and cultural exchanges. Public instruction and the university became more pervasive in the industrializing countries, and then later—in our century, that is—throughout many of the former colonial territories.

As natural philosophy gave way to biology, chemistry and physics, and of course mathematics, we learned more about the nature of matter and the substances that it forms. Innovative knowledge in the realm of science led to the invention by the head and hand of new ways to use old materials and new forms for many substances. Printing and literature, the fine arts as well as the 'domestic arts' grew, spread and became more complex.

One of the truly significant moves forward of our knowledge is the progress made in medicine and public health in only the last two generations. As recently as the 1930s, physicians were still unable to cope with most illness. If the patient had cancer or diabetes or 'heart trouble' (i.e., degenerative disease), he or she was almost invariably irretrievable from the moment of diagnosis. In about 1940 the potential of science (and, to a great extent, technology) moved biochemistry closer to physiology, microscopic histology nearer to morphology, nutrition much closer to genetics, pathology and immunology.* A unity began to appear among the life sciences, a field theretofore seriously compartmentalized. The practitioners of Greek or western medicine had had to use the force of their presence beside the patient to cure, or at least to palliate illness, when deteriorative maladies were still rampant and society lacked science-based remedies.[4]

This is not to suggest that the 1980s bear witness to a newly arrived holistic medicine, to the cure of all health anomalies. Far from it, of course. The point to be made is that, despite the frequent criticisms made of the value of contemporary science and technology, these have served instead of suborned the life-sustaining potential of epidemiology and the many other branches of clinical medicine. Some of our authors have touched on various historical aspects of the remarkably fast advances made in the health sciences since the industrial revolution took root and spread.

Some recent creativity in the sciences

In Table 2 are listed the probably most significant leaps forward made in medicine, non-human biology, chemistry, physics, the behavioural sciences, environmental control and economics during our century. I can,

---

*Indeed, specialists now view the most critical period in the growth of the human being as the perinatal period, from conception until the age of about ten months. The ravages of malnutrition of the fetus and child or of the mother (or of both), combined with inadequate midwifery or obstetrics, contribute to what some call 'internal brain drain': that portion of a country's population, which in some poor developing nations can reach 5 percent, that will always be dependent socially and physically on the rest of society.

Fig. 1.   *The woman has performed irregularly as innovator (a social activity), inevitably as a function of the cultural and economic roles she is expected to play within her family, community, educational system, and overall society.* The Woman Inventor, *whose originator and editor was one Charlotte Smith, was published in April 1891—not 1890 as shown in the cut—to mark the Patent Centennial Convention and the first century of the United States Patent Office, the Patent Office dating from 1 April 1791. A second issue of this journal, intended to be a monthly publication, has not been found. According to the National Science Foundation, women employed in science and engineering in the United States increased 157 per cent, compared to 63 per cent for men, between 1976 and 1984. In 1984, women accounted for 13 per cent of all those employed in science and engineering (a rise of 9 percent since 1976). Worldwide, women constitute on the average less than 1 per cent of the work force in engineering and science. (Photo courtesy of the Schlesinger Library, Radcliffe College, Harvard University, supplied by the Fogg Museum at Harvard; reproduced by permission.)*

Table 2. Major discoveries, inventions and other innovations of the 20th century.

Innovator	Time	Country of origin	Contribution
Karl Landsteiner	1900	Austria-Hungary	Discovery of blood types; after American naturalization, discovered rhesus factor, 1940
Karl Pearson	1900	Great Britain	Elaboration, in statistics, of the chi-squared (χ^2) test
Alfred Binet (with Théodore Simon)	1905	France	Invention of scale to measure intelligence
Albert Einstein	1905-1916	Germany	Theory of special relativity; statistical theory of brownian motion; interpretation of photoelectric effect based on Planck's quanta; theory of general relativity
Léo H. Baekeland	1906-1908	Belgium	Invention of bakelite, first synthetic resin (polymer)
Lee de Forest	1906-1907	United States	Triode tube (valve), making possible radiophony and broadcasting
George H. Shull (with D.F. Jones)	1908-1918	United States	Hybridization of maize
Ludwig Prandt, Theodor von Kárman F.W. Lanchester	1908-1920	Germany Hungary Great Britain	Explained fluid flow around a wing, founding theoretical basis of flight and supersonic flow
Raymond A. Dart	1924	Australia-South Africa	Identification of *Australopithecus africanus* in evolutionary dynamics of Hominidae
Alexander Fleming Howard Florey Ernst B. Chain	1928-1941	Great Britain United States Germany	Discovery of penicillin and establishment of its biomedical applications

Table 2.   Continued

Innovator	Time	Country of origin	Contribution
Edwin Hubble	1929	United States	The expanding universe, genesis of the 'big bang' theory
Paul Nipkow Boris A. Rosing A.A. Campbell Swinton K. Ferdinand Braun Vladimir Zworykin Philo T. Farnsworth	1885- 1938	Germany Russia Great Britain Germany Russia-USSR United States	Development of television
Russell E. Marker George Rosenkranz Hans H. Inhoffen A.J. Birch Carl Djerassi Frank D. Colton Gregory Pincus John Rock Alejandro Zaffaroni	1938- 1957	United States Switzerland Germany Australia Austria-USA United States United States United States Uruguay	Elaboration of the contraceptive 'pill'
Paul Muller Rachel Carson	1939- 1960s	Switzerland United States	Applications of DDT; origins of environmentalist movement
Alan Turing John W. Mauchly J. Presper Eckert Edward Feigenbaum Joshua Lederberg Bruce Buchanan (and many others)	1939- 1942, until 1970s	Great Britain United States United States United States United States United States	Development of the digital computer and its software; conception and elaboration of expert systems; genesis of artificial intelligence
Ernest Rutherford Frederick Soddy Irène Curie Frédéric Joliot Enrico Fermi Niels Bohr Lise Meitner Otto Hahn Fritz Strassman	1897- 1942	New Zealand Great Britain France France Italy Denmark Austria Germany Germany	Theoretical, experimental and applied development of atomic fission and nuclear fusion; development of technical basis of nuclear weapons and controlled thermonuclear fusion intended for civil energy uses

Table 2. Continued

Innovator	Time	Country of origin	Contribution
Otto R. Frisch		Austria	
Léo Szilard		Hungary	
(and many others)			
Henry Laborit	1940s-	France	Use of chlorpromazine and
Jean Delay	end of	France	lithium in order to 'empty'
Pierre Deniker	1960s	France	psychiatric hospitals
Oleh Hornykiewicz		Austria	
Arvid Carlsson		Sweden	
John Cade		Australia	
Roland Kuhn		Switzerland	
Møgens Schon		Denmark	
Oswald Avery	1944-	United States	Discovery of the double
Martha Chase	1953	United States	helix (DNA) and the genetic
Rosalind Franklin		Great Britain	code; creation of molecular
Maurice Wilkins		Great Britain	biology
Francis Crick		Great Britain	
James D. Watson		Great Britain	
and Linus Pauling		United States	
William Shockley	1945-	United States	Exploitation of solid-state
Walter Brattain	1947	United States	physics, invention of the
John Bardeen		United States	transistor, making possible modern cryptography (among others) [basic research had begun in 1930s]
Albert Einstein	1917-	Germany	Mastery of light as a
Alfred Kastler	1960	France	working instrument;
James Gordon		United States	development of optical
Herbert Zeiger		United States	pumping and the laser
Arthur Schawlow		United States	
Charles H. Townes		United States	
Theodore H. Maiman		United States	
Industrial sector	1870s-late 1940s	North America Western Europe	Evolution of research centres whose activity has highly economic goals; the coming of 'big science'

in fact, claim little credit for this. Although the table is of my own compilation, it is based on a remarkable issue of the monthly journal *Science 84* for November 1984, published by the American Association for the Advancement of Science. There are two important observations that need to be made about Table 2, for the purposes of this book.

The first is that the preponderant majority of the discoveries/inventions shown are almost never the result of the work of one individual working alone. Einstein, despite his separate listing the first time the name appears, is not an exception to this statement. We know that the preparatory work done by Mach, possibly by Poincaré, but certainly by Planck, served as inspiration for the author of relativity theory. The names of Pearson, Baekeland, de Forest and Hubble appear singly only because we do not have firm evidence of co-innovators who worked with them, or possibly elsewhere in parallel with them. This discovery syndrome beckons recall of Condorcet's remark to the effect that, when a scientist confronts a new problem, he can approach its solution by exploiting all the knowledge accumulated by his predecessors.

A second point to be made relates to the last entry on the table, the rise of centres of excellence where economic motivation is the main impetus to research carried out by teams of scientists, engineers and technicians. Edison's huge workshop at Menlo Park, employing a great number of professional inventors, may be the prototype for the research and development (R&D) group.* Edison's was the laboratory that produced the incandescent electric lamp—curiously invented at the very same time by Joseph Swan in Great Britain. Edison, it should be noted, had little formal education, understood almost no mathematics and had little real knowledge of physics and chemistry. He knew nothing of biology. Yet he accumulated 1,093 patents in his own name and left behind 3.5 million notebook pages and letters. When inspiration failed Edison, his renowned perspiration took charge, and the very nature of creativity assumed once again an individual and even personal character.[5]

Patents and copyrights are among the devices that civilisation has developed to honour and protect the creator. The recording of inventions and other innovative processes reaches back three thousand years in Egyptian, Chinese and Indian history, but the earliest known *official* recognition of an innovation (a patent) was issued as recently as 1421, in

---

*Daniel Boorstin, in *The Discoverers* (1983), prefers Prince Henry the Navigator (1394-1460) and his research centre at Sagres, Portugal, as the R&D model.

Florence. Fifty-three years later, Venice enacted the first *general* patent law. The model for western patent systems remains, however, the Statute of Patents and Monopolies of Great Britain (1623). Contrary to the tradition in China, the inventive leadership assumed by western Europe during the past five centuries may be attributable to the social rewards and personal distinctions made possible in Europe after the Renaissance by the patent system. Physical invention became recognized, therefore, as a largely individual act, leading Whitehead to claim that, by the 19th century, the greatest invention was inventing the method of invention.

Whether the result of individual or collective effort, scientific discovery and technical innovation have been shown to be correlative with the 'boom' cycles in the free-market economies by the Soviet economist, Kondratieff, more than a half-century ago. More recently, the origins of crests in the Kondratieff wave have been associated directly with surges of creation and innovation. Is this, indeed, so? As early as the Middle Ages, Ibn Khaldun, a Muslim specialist, had detected a cyclical cadence in the growth and waning of political systems—as has Arnold Toynbee during our own century. The hypothesis needs further exploration in order to be substantiated.

Creativity in the detection and memorization of sound

The origins, evolution and repercussions of innovation can be fascinating to reconstitute and trace through history. The preparation of this book coincides with the centenary of the automobile, but we shall not repeat here the debate as to whether Malandin and Delamare-Deboutteville (1885) or Carl Benz (1886) presented the world with the first truly autonomous horseless vehicle. (The French version met all the qualifications, but broke apart during its inaugural trial; the German exemplar had its trial run a year later, without incident.)

In the place of the automotive carriage, I opt for the example of electroacoustic detection and recording—the interdisciplinary system developed to preserve the human voice and its near relative, music, as an elaboration of the collective memory of mankind. Maurice Jessel, of the Mechanics and Acoustics Laboratory in Marseilles, has done this nicely. [6]

Jessel identifies the first technological step in a long chain of derivations as the phonautograph, invented by E. Scott de Martinville in 1857. Alexander Graham Bell's microphone and telephone followed 19 years later and in the following year, E. Wermer created a prototype of loudspeaker and Charles Cros devised a sort of phonograph called the palaeophone. In the same year (1877), Edison put together the tinfoil phonograph which was followed in another four years by his wax-cylinder phonograph. In 1886 Bell elaborated a 'graphophone,' and Berliner in the following year presented the first record (disc) gramophone. The recording of sound was thus launched for good, especially since by 1893 Deutsche Grammophon Gesellschaft was able to assure the duplication of records by the pressing technique.

There followed innovations that varied, improved and raised the fidelity of electroacoustic memorization. Lee de Forest (see Table 2) presented the three-electrode valve or triode in 1906, E.A. Lauste refined photographic sound-recording in 1913, and the exigencies of both civil navigation at sea and military applications during the First World War helped bring radio into the picture. In 1926, P.M. Rainer invented pulse-code modulation and the Victor Talking Machine Co. made electrical recording possible. Two years after, K. Stille assembled the first metal-ribbon magnetic recorder, while Pedersen and Poulsen perfected the photographic sound-track (making talking films possible).

In the early 1930s, AEG introduced tape-recording using iron-oxide tape and EMI made available the first stereophonic records. On the eve of the Second World War, Walt Disney Productions publicized the stereophonic sound-track for cinema, and during the war AEG developed the vinyl-ribbon tape recorder, Decca devised 'full-frequency range recording' and the Americans experimented with microgroove recordings ('V-discs' sent to the expeditionary forces). By 1948, the Bardeen-Brattain-Shockley triumvirate of innovators (Table 2) had mastered the semiconductor transistor effect. The reader is generally familiar with the stupendous progress made since that time in the recording of music and other sounds, including the Compact Disc introduced by Philips and Sony at the beginning of the 1980s, and the Japanese invention in 1986 of the digital audio tape (DAT).

During this century of invention and improvement of sound recording, two collateral developments should be added to Jessel's fascinating reconstruction. Both occurred during the 1930s, already a half-century past. The first was an intimate artistic and technical collaboration; the musician and symphony director, Leopold Stokowski, worked closely from 1930 until 1940 with sound engineers from the Bell Telephone Laboratories to augment the fidelity of recording techniques.[7] This cross-disciplinary unison of effort renders concrete J. Bronowski's observation (made some years later) that 'the act of creation is . . . the same in science as in art.'[8]

The second innovation was of an intricately technical character, when an obscure and impoverished Austrian inventor in London by the name of Paul Eisler thought up the printed circuit board (PCB) in the late 1930s. The printing of circuits on boards was intended to simplify wiring, and coincidentally, reduce considerably the costs of manufacture and maintenance. After a major British firm turned down Eisler's invention because 'girls are cheaper' on the production line, [9] the innovator had to wait for the war years until the National Bureau of Standards in the United States promoted the PCB in order to ensure a reliable method of mass production of proximity fuzes for installation in certain tactical missiles. Other than salaried income, Eisler earned no fortune from his creation—the   forerunner   of   miniaturization   and   the microchip/microprocessor. The mass production of electronic devices made it possible, by the 1970s, for Brazil, Mexico and former colonial states in Southeast Asia and the Korean peninsula to emulate North America, Europe and Japan in this important manufacturing sector.

Modernization and brute invention

Another reason to examine the genesis of innovation is because of the role of fortuity or opportunity, or merely happenstance. Two examples will suffice to illustrate this source of innovation. The first pivots about military recruitment on a systematic basis. When Louis XV was extremely displeased with his country's loss of the Seven Years' War (1756-1763), he ordered his war minister, the Duke of Choiseul, to revamp the country's entire military organization and operation. Choiseul asked J.-B. Vaquette de Gribeauval, a future general who also appears in Chapter 13 of this book in a different context, to oversee the reorganization and modernization of the royal artillery. After the redesign and trials of standardized ordnance, vehicles and drill, the new *système Gribeauval* was officially adopted in 1765 but not put into service until 1776 when the Count of Saint-Germain was named war minister under Louis XVI.

This modernization played a significant part in spreading French power, especially after the Estates General ordered in 1793 a *levee en masse* (conscription) that made France's army the strongest in Europe, combining revolutionary fervour with Napoleon's personal vision of the world. The motive of revenge against Britain (France and Britain were the 18th century's superpowers) also helped the French decide to align themselves with the rebellious British provinces of North America, furnishing the Americans some arms rendered outdated by Gribeauval's innovations. (Developing countries of the 20th century would term this the transfer of second-hand technology.)

The Prussians suffered numerous defeats as a consequence of spreading Napoleonic power, notably the wrecking of their professional soldiery at Jena in 1806. Taking stock of their plight, the Prussians re-organized, in turn, their own military system (1807-1813) under the supervision of August W.A. von Gneisenau (later field marshal) and upon the detailed recommendations of Gerhard J.D. von Scharnhorst (later lieutenant-general) and an aide named Hermann von Boyen. Reform included conscription of the majority of youths from both urban and rural areas and training them, when necessary, in reading, writing and arithmetic. Not only did this move create a reliable and substantial reserve force, it laid the groundwork for public education in Germany. This reservoir of trained manpower astonished the world by making possible its country's victory in the Franco-Prussian War and putting to route all of France in 1871.

For want of a nail . . .

Joseph Cugnot of Lorraine built the first steam-driven automotive vehicle (1769-1771) and in the process converted the rectilinear motion of pistons into continuous rotary movement. This was only a few years before the innovating chemist, Antoine Laurent de Lavoisier, was condemned to death (1794) during the Terror. In pronouncing the death sentence of the author of the law of the conservation of mass and chemical elements, J.-B. Coffinhal-Dubail said, 'The Republic has no need of savants.' Cugnot's vehicle, the progenitor of tens of millions of cars, trucks and trains, was indeed not used by either the Republic or Napoleon. Bonaparte turned down a study to adapt the Cugnot engine for the traction of artillery, delaying by several decades the advent of self-propelled transport. By 1812, on the plains of Russia, a shortage of horses and an accompanying failure of logistics forced Napoleon to beat one of history's best known retreats.[10]

The story of self-propulsion does not end in 1871, however, when France's Third Republic was formed and found itself saddled with a reparations debt worth five thousand million gold francs. This indebtedness was to feed, as it were, the spiral effect endemic to the military-industrial complex and its relentless devising of new ways and means to cope with potential adversaries and their armaments. The French goaded themselves, over the next few decades, into preparedness for any eventuality.

One of the devices sponsored by the French military during the earliest years of our century, when memories of the Franco-Prussian War were still painful, was a series of contests and prizes for novel automotive vehicles and the fast developing lighter-than-air craft. War came, and the French suffered (with their British and Russian allies) repeated reverses in 1914-1915, followed by the campaign of attrition launched at Verdun in February 1916 by German General Erich von Falkenhayn. For the French, Verdun evoked the transformation of Gaul into France, in the year 843, and later humiliation at the hands of the Prussians in 1792 and again in 1871. Although the Germans planned to bleed French forces on the Lorraine battlefield, the French opted to hold the Verdun lines at any cost.

The cost included the organization of massive road convoys along 67 km of the two-lane road leading from Bar-le-Duc to Verdun, somewhat to the northeast. This was the brainchild of a young staff captain (later general) named Aimé Doumenc. Ten thousand territorial or home-guard soldiers spread 700,000 tonnes of crushed stone and gravel in order to metal the 'Sacred Road' that would bear 3,500 Berliet CBA flat-bed trucks, 500 Renault EG gun tractors (Photos 1 and 2), and 3,000 assorted Peugeot, Latil and Saurer ambulances, liaison vehicles and utility models mounted with radios and pigeon lofts. Berliet alone would supply 25,000 lorries to the French government by 1918.

On the opposite side of the bloody front, Falkenhayn's logistics were impeded by much horse-drawn equipment and chronic insufficiencies of petroleum products and rubber. German tire material was often straw-filled leather, quickly worn out. An inflexible railway supply network was supplemented by overly heavy, steam-driven military movers that floundered on earthen lanes and in fields turned to mud. The campaign of mutual abrasion lasted through the balance of 1916, when the French finally scored a pyrrhic victory over the enemy and for which 600,000 young Germans and Frenchmen surrendered their lives.

The automotive road vehicle was joined, during the First World War, by the airplane and the tank to make the defeat of the Central Powers inevitable. Although mobile armour was of British creation, the French developed the light Renault model that, transported over long distances by imported American heavy Liberty trucks, carried the day for Allied land forces during the crucial Second Battle of the Marne (1918). Then the conflict ended, but Captain Doumenc's strategy at Verdun evolved after the hostilities into what was to become highly organized highway transport—now a civil supply mode used nearly worldwide.

Enters the versatile textile

The second example of opportunity, perhaps better known, concerns the transformation of the textile industry on both sides of the Atlantic Ocean

Photo 1. Berliet type CBA canopied flat-bed truck (1914), ordered by French Army. Power: 25-hp, 4-cylinder type Z gasoline engine, with high-tension magneto ignition. Beginning in 1914, the Berliet factory in Lyon produced 40 of these per day. During the ten-month battle of Verdun, this vehicle helped move 90,000 men, 50,000 tonnes of cargo each week. Courtesy Fondation de l'Automobile Marius Berliet.

Photo 2. Renault type EG artillery tractor (1915), with 45-hp, 4-cylinder gasoline engine, four-wheel drive, steerable front and rear axles. Reconstructed 1985-1986 from hulk found in the Vosges mountains. Courtesy Fondation de l'Automobile Marius Berliet.

and, later, elsewhere. Research work on nylon was undertaken in 1927 by DuPont and Imperial Chemical Industries. The rise in the cost of such natural fibres as cotton and silk, the advent of the electromechanical washing machine for home use, and the gradual disappearance in industrialized countries after the First World War of a servant class pushed chemists to search for new fibres. They sought synthetic threads resembling good-quality wool or mohair, silk and even linen and flannel. Research lasted almost 12 years, whereupon the first non-aromatic polyamides were successfully synthesized and placed on the market in 1938. Nylon, derived from a combination of the names New York and London, is resistant to bacterial and fungal action and to attacks by insects and rodents; natural fibres are not. Nylon is solid, tough, can be moulded, drawn and used as parts in the assembly of electrical apparatus. Cast as bearings or gears, nylon requires little or no lubrication and functions relatively silently.

The opportunity factor and the boundaries of knowledge

The decade of the 1880s, a century ago, was a period of intensive movement ahead in innovation of many kinds. The telegraph already circled the globe; the electric lamp and telephone had just been introduced, the automobile was born in 1885-1886, and radio was not far behind. All of these symbolized the second step on the staircase of the Industrial Revolution, and as inventions they were destined to change the life of mankind almost everywhere on the planet. In Japan, society was making a jump from feudalism directly to industrialization. It is curious to note, too, that Hitler, Mussolini, Tojo and Stalin were all born within the decade of 1879-1889: these were the 'innovators' of oppressive social processes of a magnitude previously not experienced by mankind.

Einstein had just been born (1879), Charles Darwin and Gregor Mendel had just died (1882, 1884) and Niels Bohr was born in 1885 in Copenhagen. The natural sciences were passing through a critical phase of evolution, as Camille Flammarion's *Astronomie Populaire* (1879) also bore witness during this exciting decade in the development of perception and analysis.

Curious are a number of factors in the story of two near-contemporaries of the time, Charles Darwin and Johann Gregori Mendel. The first remains to our day a paragon of the descriptive investigator, while the second has gone down in history as the model experimental researcher.

First, we note that neither was a particularly diligent (or even interested) student. Yet the two became outstanding thinkers of their time, leaving forever their mark on the history and philosophy of biological science. Poor performance in school (as was to be the case later with Einstein, too) did not curb the natural abilities of Darwin and Mendel to 'ask the right questions,' to investigate, then compare and contrast and further analyze their respective work before finally making public their astonishing findings in a modest way.

Secondly, each took advantage of the opportunities at hand in order to push a methodical examination of the nature around them—one aboard a naval vessel roaming the South Atlantic and Pacific Oceans, the other within the confines of a monastery garden in Brno of impressively limited proportions. Both Mendel and Darwin exploited expediency to the fullest, therefore, and the results of their endeavour justified their way of doing the most with the means available.

A third observation concerns the interest manifested by Darwin and Mendel in fields essentially alien to what they had known and experienced earlier in life. Their minds were open, that is to say: they were open to innovation in understanding, in knowing. The novelty of their experiences, furthermore, propelled Mendel and Darwin further and deeper in their investigations. The two were, in a word, pioneers; and as pathfinders, Darwin, Mendel and some of their contemporaries were not afraid of change or adaptation to change. Indeed, they did not fear the adoption of entirely new attitudes and convictions concerning man's knowledge of nature and how it works.

A fourth and final impression is intimately linked with the previous one. Mendel and Darwin, by exploring new limits of knowledge through a quantum leap, realized that they stood on the threshold of a possible evolution in civilisation itself. Their work, unique by its very nature, was not however the only novelty in a new age. Faraday's findings in electromagnetism, Maxwell's in electrodynamics, Carnot's in thermodynamics, the establishment by Mendeleyev of the periodic table of the elements, new discoveries in geology, climatology and astrophysics, the continuing spread of the Industrial Revolution—all this diffusion of innovative knowledge and methodology was happening simultaneously, intensively. Much of the world was on the verge of leaving the age of iron and steam to enter the epoch of electricity and petroleum before advancing further to today's era of the atomic nucleus, microelectronics and expert systems. One may conclude that our present and future knowledge—our innovative overtures—continues to be formed in great part by the fortuitous but persistent spirit of Darwin, Mendel and their fellow-thinkers of ten short decades ago.

Summary of the probable mechanics of inventivity

Can rhyme or reason be extracted, or at least interpreted, from the foregoing? We seem to have considered, in effect, four major sources of innovative knowledge and the creative spirit.

- *Inspiration*, as in the instances shown on Table 1 and reinforced by the example of Edison and the persistent inventors. Painters, sculptors, poets, composers and design (as opposed to engineering) architects occupy this category. Some analysts call inspiration 'imagination' or the 'rapture of youth' (since it is usually experienced before middle age).

- *Investigation*, applying not only to most of the personalities listed in Table 2 but also to patient researchers in the human sciences: sociology, history, economics, anthropology palaeontology, psychology, ethology and ethnology—to name but a few; the physical sciences related to the earth and the solar system; man's vestimentary and gastronomic habits; nutrition; and veterinary medicine. Both investigation and the following category are sometimes classified as 'experience.'

- *Experimentation*, easily confused with the previous classification, excludes such original thinkers as Newton and Einstein who limited themselves to 'thinking problems.' The category includes craftsmen and other artisans using empirical technology, the result of generations or even centuries of trial-and-error activity in all fields of culture. (Culture = what head and hands—and sometimes the feet, as in dance and championship sports—do.)

- *Search for economic niche*, of which the illustrations of nylon and the printed circuit board are archetypes. This is the impetus behind much pharmaceutical research, too, a field in which some investigation and development are ruled out *a priori* because of the 'uneconomic' nature of certain rare diseases. The artist and writer are often motivated in this way, and in industrial R&D the device of goal-setting can be instrumental in focusing concentration on innovation.

Allusion has already been made to failed innovation: creation that missed getting under way. The story of mankind is replete with writers, artists and musicians who failed to be recognized, with scientists and philosophers who have not won the highest accolades (why have descriptive investigators—palaeomorphologists and evolutionists—not been attributed Nobel or Lenin Prizes?), with engineers and builders whose edifices fell to pieces or exploded, with designers of vehicles or missiles that failed to perform correctly.

Did the roof of Beauvais Cathedral, for instance, collapse in 1284 because of the 'Gothic' masons' inability to comprehend the mathematics of stress and strain, or was their workmanship inadequate? Postal networks, to take another example, are relatively new to civilisation, but some of them are showing unmistakable signs of system erosion. Was the polar explorer Ernest Shackleton an inadequate technician and a poorer leader, or was his quest for the South Pole a monumental (and tragic) exercise in masochism?[12] The Maginot Line, built of fearful optimism and at great cost, and the Japanese *kamikaze* or *shimpu* corps born of patriotic despair as well as demanding the highest sacrifice, were two war-related innovations of our century that proved to be irreversibly ill-fated.

Failed commercial products and especially services are legion, contributing directly to the obituaries of small and medium business enterprises. They are often novelties that would not or could not 'work' on the market. Yet the genesis of any defective innovation can be classified in one or more of the sources of the creative spirit cited earlier.[13] See Figure 3 for another observer's view of the innovative processes, and note here the element of history. History occupies a position of past in relation to the psychology and philosophy of innovation, yet its relation to economics and the political milieu implies the present and (perhaps more importantly) the future.

One of the greatest Greek inventions was the idea of history. The word 'history,' along with its cognates in European languages, derives through the Latin *historia* from the word *historiê*, which the Greeks used to mean 'inquiry' or 'knowing by inquiry.' Its original meaning survives in the expression 'natural history' for inquiry into nature.[14]

Thus irrespective of the uses to which we put the terms history, creativity and invention, all three signify change: past, present, and future. ■

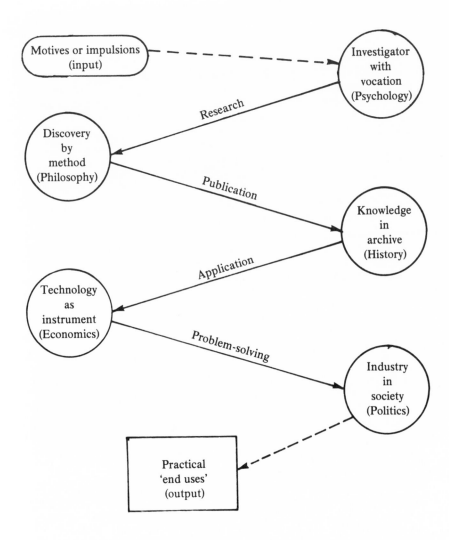

Figure 3. Schematic plan of one specialist's view of discovery, especially in terms of the conversion of 'pure' science to its ultimate industrial possibilities as product, process, design or service. Adapted from John Ziman, An Introduction to Science Studies, The Philosophical and Social Aspects of Science and Technology, *Cambridge, Cambridge University Press, 1984, p. 29. If patent or other proprietary interests are involved, the publication step is often omitted.*

Notes

1. J. Pelikan, *The Mystery of Continuity*, Charlottesville, The University of Virginia Press, 1986.
2. Stephen Jay Gould has most effectively propagated, throughout his many publications, the notion of punctuated equilibrium.
3. See the report, Unity of Biology, by F.B. Straub of the Hungarian Academy of Sciences, in: *Biology International*, No. 12, December 1985, p. 2.
4. *Cf.* E. Shorter, *Bedside Manners: The Troubled History of Doctors and Patients*, New York, Simon and Schuster, 1985; and S.J. Reiser, The Emergence of Scientific Medicine: A View from the Bedside, in: C.G. Bernhard, E. Crawford, P. Sorbom (eds.), *Science, Technology and Society in the Time of Alfred Nobel*, Oxford, Pergamon Press, 1982, p. 121.
5. There is a fair amount of biographic literature available on Thomas Edison. One of the most complete and objective works is the effort of a Briton, Ronald W. Clark: *Edison, The Man Who Made the Future Work*, New York, G.P. Putnam's Sons, 1977. Clark specifies one of the motivating impulses of the professional inventor as 'the essential spur: the knowledge of what was wanted' (p. 29).

   What Edison is supposed to have said to generate the inspiration-perspiration legend was in response to a question raised by his secretary, Samuel Insull, as to what constitutes genius. Edison replied, '. . . about 99 percent of it is knowledge of the things that will not work. The other one percent may be genius, but the only way that I know to accomplish anything is everlastingly to keep working with patient observation.' Clark, *op. cit.*, p. 89. (Clark is the biographer of two other innovators, Einstein and Bertrand Russell, and the author of *Works of Man*, New York, Viking, 1985.) See also F.L. Dyer, T.C. Martin, W.H. Meadowcroft, *Edison, His Life and Inventions*, New York and London, Harper & Brothers, 1929.
6. M. Jessel, Enregistrement et reproduction du son, *Impact: science et societe*, No. 138/139, 1984, p. 93; available also in English as Sound Recording and Reproduction (p. 91), and in Arabic, Chinese, Korean and Russian versions.
7. R.E. McGinn, Stokowski and the Bell Telephone Laboratories: Collaboration in the Development of High-Fidelity Sound Reproduction, *Technology and Culture*, Vol. 24, No. 1, 1983, p. 38. Also worth noting are Margaret Cheney, *Tesla: Man Out of Time*, Englewood Cliffs, Prentice-Hall, 1981, and F.B. Jueneman, The Sound of Music, *Research & Development*, February 1986.
8. J. Bronowski, *A Sense of the Future* (P. Arioti, R. Bronowski, eds.), Cambridge, Mass., MIT Press, 1977.
9. B. Fox, Father of the Third Revolution, *New Scientist*, 30 January 1986.
10. J. Richardson, L'innovation et le progrès, *Encyclopédie Clartés* (coll. Métiers 1-3), Sciences et Techniques Actuelles (Nos. 3 and 4), October 1985, p. 7002-7; see also N. Rosenberg, The Growing Role of Science in the Innovation Process, In: C.G. Bernhard, E. Crawford, P. Sörbom (eds.), *op. cit.*

11. ————, Boundaries of Knowledge: Opportunity as Prologue, Symposium on Science and the Boundaries of Knowledge—The Prologue of Our Cultural Past, Venice, 3-7 March 1986. See also S.J. Gould, Evolution and the Triumph of Homology, or Why History Matters, *American Scientist,* Vol. 74, No. 1, 1986, p. 60.
12. R. Huntford, *Shackleton,* New York, Atheneum, 1986.
13. For a fascinating account of social heroics gone wrong during almost a millennium of Japanese history, see I. Morris, *The Nobility of Failure,* New York, Holt Rinehart Winston, 1975.
14. D.J. Boorstin, *The Discoverers,* New York, Random House, 1983, p. 562.

To delve more deeply

ANDERSON, H.H. (ed.), *Creativity and Its Cultivation,* New York, Harper & Row.

BEARD, G., *Legal Responsibility in Old Age,* New York, Russell, 1874.

DAWKINS, R., *The Blind Watchmaker,* New York, W.W. Norton, 1986.

GARDNER, H., *Frames of Mind: The Theory of Multiple Intelligences,* New York, Basic Books, 1983.

GELPI, E., Les jeunes: leurs créations et leurs contraintes, *International Review of Education,* Vol. XXXI, 1985, p. 429.

KAGAN, J. (ed.), *Creativity and Learning,* Boston, Mifflin Company, 1967.

KOESTLER, A., *The Act of Creation,* New York, Macmillan, 1964; London, Hutchinson, 1964.

MAYER, R.E., *Thinking, Problem Solving, Cognition,* New York-San Francisco, W.H. Freeman, 1983.

MAYOR ZARAGOZA, F., La investigación científica como creación, In: J. Arana (ed.), *Creatividad Mediterránea,* Madrid, Fundación Instituto de Ciencias del Hombre, 1983.

MORRIS, R., *Time's Arrow,*New York, Simon and Schuster, 1985; see esp. chapter 5.

PFEIFFER, J.S., *The Creative Explosion, An Inquiry into the Origins of Art and Religion,* New York, Harper & Row, 1982.

ROBERTSON, A., Technological Innovations and Their Social Impacts (introduction to special issue on 'Technology and Cultural Values'), *Intern. Social Science J.,* Vol. XXXIII, No. 3, 1981.

ROOT-BERNSTEIN, R.S., Creative Process as a Unifying Theme of Human Cultures, *Daedalus.* Summer 1984, p. 197.

SIMONTON, D.K., *Genius, Creativity, and Leadership: Historiometric Inquiries,* Cambridge, Mass., Harvard University Press, 1984.

SMITH, M.R., *Military Enterprise and Technological Change,* Cambridge, Mass., MIT Press, 1985.

SUCHODOLSKI, B., *Permanent Education and Creativity*; KUCZYNSKI, J., *Creativity as a Practical Philosophy* (trans. C. Fletcher) (doc. ED-82/WS/16, Paris, Unesco, January 1982.

VANDERBURG, W.H., *The Growth of Minds and Cultures* (with foreword by Jacques Ellul), Toronto, University of Toronto Press, 1985.

WARNIER, J.-D., *L'Homme face à l'intelligence artificielle*, Paris, Editions d'Organisation, 1984; *Computers and Human Intelligence*, Englewood Cliffs, Prentice-Hall, 1986.
WEISBERG, R.W., *Creativity, Genius and Other Myths*, New York, W. H. Freeman, 1986.

The Journal of Creative Behavior (any issue), a quarterly published at 1300 Elmwood Avenue, Buffalo, NY 14222 (United States) in the tradition J.P. Guilford and his profound interest in the manifold possibilities of human intelligence.
The corporate-business reader will be interested in the activities of the Center for Creative Leadership, 5000 Laurinda Drive, Post Office Box P-1, Greensboro, NC 27402, tel: 1-919-288-7210, and in its quarterly magazine, *Issues & Observations*. The same reader may wish to consult *Creativity and Innovation Network*, 'an international forum for all those concerned with the discovery processes,' available from: Marketing/Subscription Manager, Room 3.1, MBS, University of Manchester, Booth Street West, Manchester M15 6PB (Great Britian).

Contemporary life is experiencing its third major technological upheaval in two centuries. Industrialized countries are being deeply affected by the computerization of society, a phenomenon that is even changing basic value-systems; its effects are slowly reverberating in the developing countries. Everywhere, the processes of innovation are also changing because of the evolving nature of ambient research, technology and socio-cultural systems. Perhaps even more significantly, the poles of invention and social change are shifting too.

Chapter 25

The evolving concepts of creativity and invention (Postface)

Junnosuke Kishida

The author of this final chapter was trained as an aeronautical engineer at the former Tokyo Imperial University, but after the Second World War turned to a career of journalism and social criticism. He retired in 1985 as chief editorial writer of the prestigious daily Asahi Shimbun, *a major Tokyo and national daily newspaper. Mr. Kishida is the author of several books dealing with the science-society interface. He is currently Honorary Chairman, Japan Research Institute, The Forum, 4-1 Kioi-cho, Chiyoda-ku, Tokyo, Japan.*

Man's newest industrial revolution

The industrialized nations are at the threshold of what some call the third industrial revolution. Man experienced the first of these in the last quarter of the eighteenth century. The invention of the steam engine and its application to various systems, especially spinning machines, introduced serial or mass production. Thenceforth, processes for working with various materials (especially iron) improved constantly: this was the first industrial revolution.

About a hundred years later, during the 1870s, industry entered a new phase, characterized by the harnessing of electrical energy. The merits of this new source of power—especially its fast transmission and flexibility of use—gave it a key role in the marvelous industrial and other technical progress launched around the 1930s.

But the good in life is usually accompanied by the bad. As this revolution took form, pollution, destruction of the natural environment, and waste of energy and other natural resources became more serious; it led finally to the petroleum crisis of the 1970s. All of this has forced us to pay much attention to the maintenance of the proper conditions of life, to economize on natural resources and energy, and to try to transform our industrial structure. As we accomplish this (more or less), we find ourselves faced with a new industrial revolution.

The revolution is further refined

The rapid pace in today's change, as we know, is in the field of information processing: novelty that is related to computer technology and communications. The essence of the newest revolution is the technics that might substitute for the brain. This is in marked contrast to what transpired as a result of the first industrial revolutions, in which machines replaced human brawn in the handling of heavy loads and new means of transport broadened mankind's geographic horizons considerably.

Today's exploitation of basic materials and of infinitesimally smaller components, high in quality and low in cost, has made it possible for many people to have access to a word processor or to a computer. Worldwide telecommunications have been greatly facilitated by the advent of satellites and lasers. This is why (in the 'advanced' countries, at least), yesterday's claim to an industrial society gives way easily to that of the 'informatised' society typifying the latest industrial revolution.

Emulating brain and body

Our concerns today are increasingly related to notions such as physical interconnection, merger and synthesis. With the first progressive steps of industrialization, there was a continuous effort towards 'bigger and better,' which iron-and-steel and electrical industries exemplified. Their motto might well have been, 'Heavy, Huge, Long and Broad.' In today's computerized world, however, what seems to count is 'Light, Small, Compact and Fine.' In many fields, semiconductors among them, there is rapid miniaturization and overall improvement in quality.

The new philosophy of smallness is the key both to change in the industrial structure and reduction of our rate of consumption of energy and natural resources. If our 'heavy' industrial sectors try to rationalize their organisations, really modernizing in order to renew themselves, there is no alternative but to find a way to introduce 'smaller is better' to their heavy output and their immense scales of production. It would be quite natural for heavy industry to find a way out by combining and merging 'small' with 'heavy' because such a move is reminiscent of the relationship between the brain and the body's other organs.

Combination and functional merger apply to other spheres as well. Take, for example, some of the pioneering work being done with robots. These are the stars of the new field of robotics, combining mechanics with electronics.

Research on communication via laser is generating keen competition among various countries interested in the new possibilities of lumino-electronics, a wedding of the laser with electronics more generally. This technology is exploring virgin fields in combination with glass fibres, expanding broadly the area of fibre optics.

New integrative processes

Electronics have also married with the biotechnologies and medicine to form the fairly new world of bionics which, to a certain extent, is one of the formative zones for artificial intelligence. Similarly, the gradual alliance of the computer with telecommunication systems has given birth to teleprocessing or 'telematics,' intended not only to speed a computer's output but to raise the efficiency of communications in general.

It is true that computerized society is fundamentally an improved 'telecommunicative' society whose main bulwark is the process of synthesis. This and the other examples that I have given attest to the specificity of today's consciousness of information processing; this is the third technological revolution, therefore, that of new physical make-up, interconnection, merger and synthesis. We can expect such specificity to put a new face on the idea of creativity, and blazing innovative pathways to the synthesizing process should require means other than the traditional.

The very concepts of the creative spirit and of invention are, indeed, being transformed.

Our age even insists upon new kinds of specialists, those who can meld their efforts with those of others. Today's and tomorrow's innovators not only need extensive knowledge and a deep interest in their own disciplines but in many of those seemingly of only incidental interest, as well. The new innovator has to be able and accustomed to combine areas of knowledge which heretofore were considered essentially separate and discrete.

This evolution of the creative processes confirms that the collegial approach to investigation—often referred to throughout the preceding chapters of this volume—is in itself a creative and inventive approach. One needs to appreciate, too, that this novel creative spirit (although not originating in the profound meditation of some genius working alone in his laboratory) is not something of rare precocity.

Some of the limits of research

In order to make varied creativity more ubiquitous than in the past, the free exchange of information is mandatory. One would hope to eliminate obstacles to such exchange, military secrets, for example. Industrialized countries lacking huge military power are particularly favoured in this regard, nations such as Japan.

The cradle of creativity and where it spreads are not limited, it must be added, by developments in high technology. In an affluent society, the public's requirements have no bounds. In responding to new requirements, the creative processes also take multiple forms, and the constraints inherent in military technology are far less confining. Civil technology, responding constantly to new pressures put upon it, plays well the role of the mother of invention.

Originality and pragmatism in Japan

After the Second World War, Japan introduced with vigor the advanced technologies of the most progressive countries, and it did so in such a systematic way that today Japan is competing with the most advanced of these countries, and in a number of fields. Yet when it comes to true creativity, as it is sometimes said, Japan is inclined to lag behind. The nation benefits splendidly from applications, but its efforts in the way of basic research remain inadequate.

Having said this, I cannot overlook the fact that the Japanese economy—responding to consumer demand—has developed designs and goods highly useful in terms of safety, security, facility of processing or maintenance, and low cost. And, occasionally, Japan has brought forth entirely new products and designs.

A new world of multiple centralization

There are daily new opportunities to forecast that the future will belong to Asia and the Pacific Ocean basin. This does not mean, of course, that the other regions will experience a reduced role. Until now, however, the 'hub of the universe' has been western Europe and the United States. In the future, there will be several such hubs, so that one can easily expect Asia and the Pacific to become one of these key regions.

Computerized society already attests to diversification, whereas a computerized world will bear witness to 'multiple centralization.' And in such a new world, all nations will be able to stake a claim, and to recognize mutually, their just values derived from their own history and tradition.

Mankind ceaselessly effects, of its own impetus, evolution in culture and civilisation that are quite different from the evolution of living cells. If we could use a single word to describe this evolutionary process, this continuous advancement, it is towards *diversification*.

Creativity itself is diversifying. In our acceptance of physical make-up, interconnection, merger and synthesis, we perceive that discovery and invention have variants. So, judging by the old rules no longer applies. Our passwords to the future must be 'broad tolerance' and 'an open mind.'   ■

Translated from the original Japanese to French by Ms. Makiko Ueda.

INDEX